W. K. V. GALE

THE IRON AND STEEL INDUSTRY

A Dictionary of Terms

DAVID & CHARLES *Newton Abbot*

ISBN 0 7153 5302 0

Set on the IBM Selectric Composer in 11 on 13pt Press Roman
and printed in Great Britain by
Redwood Press Limited London and Trowbridge
for David & Charles (Publishers) Limited
South Devon House Newton Abbot Devon

A/669. 1 23

THE IRON AND STEEL INDUSTRY

A Dictionary

Contents

Preface

Like other industries which have developed from craft origins through empirical practices to its present science-based form, iron and steel making has acquired a complicated terminology of its own. It has changed over the years and is changing still. Many terms, including some current in recent years, are now obsolete and new ones have found their way into the language of iron and steel makers. Some have changed their meanings and a few which logically ought to be obsolete stay obstinately with us to describe things very different from what they originally meant.

The writer, in the course of over forty years in close connection with the iron and steel industry and in studying its technical history, has found it necessary to know the meanings of a large number of specialized terms. These, over 3,000 in all, are set out in the pages which follow.

It is not suggested that the expert should look in this dictionary for the meanings of terms within his own particular field; indeed it would be presumptuous so to suggest. But it is one of the problems of modern life that because manufacturing processes are now so complex, an expert's field becomes ever narrower, while at the same time his need for at least some knowledge of what his colleagues in other disciplines are doing is increasing. So it is hoped that the present book will help in the constant cross-referencing, from process to process and from old to new, that is becoming more necessary as time goes on.

Then there is the general reader who without specialized knowledge of iron and steel is interested in some, many, or even all, of its aspects. There is also the student of industrial history or industrial archaeology. Such a student may not even be dealing specifically with iron or steel but he is likely sooner or later to come across some of the terms peculiar to the industry. And as leisure becomes a more important factor in life this group of people is growing.

So it is to the non-specialist, to the general reader, and to the specialist who seeks a quick definition in the language of other specialized fields, that this dictionary is directed.

It should be pointed out that it is, in fact a dictionary; it is neither an encyclopaedia nor a textbook. Readers who seek more detailed information are referred to textbooks. It is biased, deliberately, to the shop floor rather than to the laboratory, for it is in this field that there is the greatest lack of published information.

There is not and never has been any central authority devoting its energies to the standardization of iron and steel terms. There are a few British Standard glossaries, but they cover only a fraction of the terms in use and they do not include those which were obsolete at the time of compilation. That is all. The definitions given here are therefore those commonly accepted in the various branches of the industry where they are, or were, in general use.

It must be added, however, that agreement is not always universal. On the whole it is, but in a few cases a definition is rather loose and there may be more than one view as to its exact meaning. The dividing line between plates and sheets, for example, is generally agreed, now, to be 3 mm ($\frac{1}{8}$ in), but into which category a 3 mm flat-rolled product falls seems to be a matter for personal decision. Some would have sheets as 3 mm and below; others say that plates are 3 mm and upwards in thickness. But there are not many such cases and they are noted as they occur.

The nomenclature of the iron and steel industry in the English-speaking world is predominantly that of the British Isles, though there are local differences. This is especially so as between Britain and the USA, but even here the variations are generally no more than are to be found in normal speech and writing. Thus the British mould is an American mold, armour plate in Britain is armor plate in America, and a melting shop in Britain is a melt shop in the USA.

There are, of course, a few words of purely local significance to certain areas of Britain which must be unknown abroad; these are noted as local where they occur. But they are not very numerous and in general the variations in spelling and in meaning between different countries using the English language are not of great importance.

A word of caution is needed. Some iron and steel terms are used in other industries and in some iron and steel *working,* as distinct from *making,* processes; here their meanings may be quite different. Thus billet in iron and steel making has one meaning, in hot-worked tube making another and in the brass trade yet another. And to the countryman, billet means none of these things. The terms as defined here are solely as used in the manufacture of iron and steel from the ore to the finished cast, forged, rolled, or drawn product.

A few of the terms included are (or were) geographically restricted; where this is so the fact is noted in brackets. Thus bellman was found only in Scotland; a boster was peculiar to the British Black Country. Purely local dialect variants of terms are not included.

Many of the terms listed are now obsolete and this fact is recorded by the note (obs). It should be explained that where a term is described as obsolete it means that it is either literally so — as charcoal iron — or that it is for all practical purposes so — as puddled wrought iron. The parts of speech are only noted as (a), (n), or (v) where this distinction is not otherwise apparent, as in: use (n).

Finally, let it be admitted that this book is not definitive. No technical book ever is. There is probably no living person who knows all the terms which are or have been used in iron and steel making. Some hitherto unknown terms will turn up as the study of history proceeds and many new ones will come into being as technology develops. But it is believed that the book is as comprehensive as it is possible at present to make it.

W. K. V. Gale

A

Abrasive cutter   An abrasive disc, rotating at high speed, used to cut steel bars. Not common in steelworks but can be used to cut hard alloys and steels.

Acetylene   An inflammable gas, a compound of carbon and hydrogen (C_2H_2) used with oxygen to provide a very hot flame for cutting and dressing steel. Not so much used now — other gases such as petroleum or natural are often used instead.

Acicular cast iron   A form of cast iron with an acicular or needle-like microstructure.

Acid Bessemer process   *See* Bessemer process.

Acid open-hearth process   *See* open hearth furnace.

Acid process   In a restricted sense any steelmaking process in which the furnace lining and the slags are chemically acid (eg silica). It will not remove sulphur or phosphorus from the iron. Is less important now than formerly.

Acid refractories   Refractories such as silica, ganister, firebrick, used for lining acid furnaces. Cf. basic refractories, neutral refractories.

Acid steel   Steel made by any process in which the chemistry is acid. Only low-phosphorus irons can be used and the process is now uncommon. The furnace linings are siliceous, eg silica, ganister.

Active mixer   *See* mixer.

Acute angles   Rolled angles in which the included angle of the two legs is less than a right angle.

Addie process (obs)   *See* Langloan process.

Addition elements   *See* Additions.

Additions (Addition elements)   Alloying elements added to a steelmaking charge, in the furnace or in the ladle, to produce the required specification.

Admiralty pattern   *See* Chequer plates.

Aerator   A machine used in a foundry for shaking up, cooling and aerating hot, used moulding sand.

After blow (obs)   The final stage in the basic Bessemer process, when the carbon has been removed but the blast is kept on for a few minutes to burn out most of the phosphorus.

AGC   *See* Automatic gauge control.

Age hardening   *See* Ageing.

Ageing (age hardening)   Some steels gain strength by internal structural changes over a period. The change can be speeded up by heat treatment (artificial ageing).

Agglomeration   Collecting together into lump form of fine iron ores, dust, and other materials which could not otherwise be smelted. Sintering (qv) and pelletizing or balling (qv) are the common processes used today.

Air belt (air box)   The windbox of a cupola.

Air box   *See* Air belt.

Air consumption   It is a little-known fact that air is the heaviest of all the materials (coke — ore — limestone — air) which go into a blast furnace. From four to five tons of air are needed for every ton of iron produced.

Aired bars (obs)   Bars of blister steel which have become decarburized through leakage of air into the cementation pots, and are thus defective. They have scale on the surface instead of the characteristic blisters. *See* Cementation.

Air furnace   A reverberatory furnace similar in shape to a puddling furnace but usually much larger. Used for melting down cast iron before the introduction of the cupola (qv) and since for melting large pieces (eg scrap rolls) or where a large quantity of molten iron is needed for a heavy casting.

Air hammer   *See* Pneumatic hammer.

Air-hardening steel   An alloy tool steel which will harden itself, after heating to the necessary temperature, simply by cooling it in air. Distinct from most hardenable steels which need quenching rapidly in oil or water.

Air knife   A device which blows a thin stream of air or steam on to the sheet surface in a continuous hot galvanizing line to remove surplus zinc and to control the thickness.

Airless shot blasting   *See* Wheelabrator.

Air port *See* Port.

Ajax process A steelmaking process developed at Appleby-Frodingham Steelworks. An open-hearth furnace is modified by fitting oxygen lances so that oxygen can be blown in under strict control at the appropriate point in the melt.

Alexander & M'Cosh process (obs) *See* Gartsherrie process.

All-basic furnace At one time open-hearth furnaces usually had an acid refractory roof even if the working lining was basic. The introduction of the basic roof gave rise to the term all-basic.

Alligator shear (obs) *See* Crocodile shear.

Alligator squeezer (obs) *See* Squeezer.

All-mine iron (obs) Pig iron made solely from iron ore, as distinct from cinder pig (qv). The term was confined to the Black Country.

Alloy Any mixture of two or more metals which mix together when molten and do not separate on cooling. Steels are alloys of iron and other elements eg carbon steel = iron + carbon, stainless steel = iron + nickel + chromium. All commercial alloys contain at least traces of other elements which are not there intentionally.

Alloy cast iron Cast iron to which other elements have been added deliberately to impart desired properties.

Alloy elements *See* Additions.

Alloy iron *See* Alloy cast iron.

Alloy steel (1) In general terms any steel other than carbon steel, to which other elements have been added deliberately to impart required properties.

Alloy steel (2) For statistical purposes the industry classifies as alloy steel any steel containing by weight at least 0.1% molybdenum, tungsten, or vanadium; or 0.4% chromium or nickel; or 10% manganese.

Alumina Aluminium oxide Al_2O_3. Used in refractories.

Aluminium (in US aluminum) A metallic element, symbol Al, used as a deoxidant in steelmaking and for surface protection of rolled steel.

Aluminizing Coating steel strip or sheet with aluminium as a corrosion inhibitor.

Ancony (obs) *See* Loop.

Angle iron (L-iron - rare) Iron or steel rolled to a right angle cross section like a letter L, or, less commonly to an obtuse or equal angle or equal-leg angle) or unequal (or unequal angle). Angle iron is obsolete but the term is often wrongly applied to rolled steel angle (RSA).

Angus Smith's composition   A protective finish based on coal tar, tallow, resin, and quicklime, formerly used extensively for protecting iron or steel pipes against corrosion.

Anneal (1)   Heating steel to a suitable temperature, depending on the composition, holding it at this temperature for the necessary period and cooling it at the correct rate, to remove stresses set up by previous processing and so soften it and/or improve its machinability and cold-working properties.

Anneal (2) (obs) ('neal') ('nail')   To heat the crucibles, over a coke fire in the annealing ('nealing' or 'nailing') grate, to prepare them for steelmaking.

Anneal (3)   To heat-treat castings of special white iron to convert them to malleable cast iron (qv).

Annealing can   *See* Annealing pot.

Annealing furnace   A furnace – it can be of various types and sizes – solely for annealing.

Annealing pot (annealing can)   A cast-iron container in which iron castings are packed for protection during annealing, especially for making malleable cast iron (qv).

Anthracite   The purest and highest grade of coal – almost pure carbon. It is used for recarburizing (qv) and was formerly used in a few places as blast furnace fuel.

Anthracite furnace (obs)   In a few blast furnaces in Wales the principal fuel was anthracite.

Anti-pipings   *See* Hot topping (1).

Anti-slip plates (non-slip plates) (padded plates) (warted plates)   Steel plates with patterns rolled on the surface for use as flooring in steel structures. Chequer plates are the most common but various other patterns can be found. These include warted plate (with little bumps on the surface at intervals), padded plates (with small projecting pads at intervals). Super Tread, Non-Slip, and Super Grip are proprietary British patterns.

Anvil (1)   The specially-shaped steel block on which a smith makes forgings. Sizes and shapes vary but the common one has a pointed end or bick (beak), a square end, and a flat surface between. A square hole (the hardie hole) in the heel accommodates various tools and a second, round hole near the hardie hole (the pritchell hole) is used for punching.

Anvil (2)   That portion of the bed of a power hammer on which the work is placed to be struck by the hammer.

Anvil pallet   *See* Pallet.

Approach table   A roller table for moving long pieces to a rolling mill, eg blooms to a billet mill. It is similar to and may be identical with, an entry table (qv) but is usually longer.

Arc furnace (electric arc furnace)   Any electric furnace in which the heat is generated between an electrode or electrodes and the charge. The electrode may be of carbon or of the metal being melted as in consumable electrode melting (qv). In an indirect arc furnace (less common) the arc is struck between two electrodes above the metal bath, which is heated by radiation.

Arch brick   A refractory brick with two faces inclined towards each other so that a number of bricks can be built into an arch.

Arch-form hammer   The most common form of steam hammer, with the steam cylinder mounted vertically on the top of a cast iron arch over the anvil.

Arcos electroslag process   An American modification of electroslag refining (qv) in which thin steel strip and alloying powder are fed continuously into the molten slag instead of a specially made ingot.

Argillaceous iron ores (Clay ironstones)   Ores in which the main waste or gangue material is clay.

Argon   A natural inert gas used for inert atmospheres in furnaces for purging, and for stirring in degassing.

Argon-Oxygen process   A steelmaking process of American origin, using an electric arc furnace for melting and a converter-type vessel for refining. Decarburizing is done in the vessel by a mixture of argon and oxygen blown through tuyeres near the bottom. Final composition adjustment is also carried out in the vessel. It is new and so far only in very limited use.

Argon teeming   Pouring steel through a blanket of argon gas into a mould or moulds filled with the same gas.

Arisings (own arisings)   Scrap, such as crop ends or faulty

11

material arising in the works where it is re-used as melting scrap. When the material comes from a foundry it is usually called foundry returns, or returns.

Armco   An American proprietary name derived from American Rolling Mill Company, used for a commercially pure iron, usually rolled into sheets. Total impurities less than 0.1%

Armouring (throat armour)   Metal pieces let into the refractory brickwork at the throat of a blast furnace, to protect it from damage by the charge.

Armour plate (bullet-proof plate)   Heavy plates, formerly of iron but now of alloy steel, specially made for naval and military purposes. Lighter plates, known as bullet-proof, are another form of armour.

Artificial ageing   *See* Ageing.

As cast   Iron or steel castings which have not been heat treated.

As drawn   Cold drawn steel not heat treated.

ASEA-SKF process   A Swedish method of vacuum degassing steel using a special ladle, a vacuum degassing canopy and an electric arc lid, either of which can be swung over the ladle. The arc lid is used to heat up the steel after degassing.

As forged   Forgings which have not been heat treated or machined. Cf black forgings.

As rolled   Any rolled steel products just as they have left the rolling mill.

Assorting (obs)   Simply the sorting, by hand, of tinplates into grades.

Aston-Byers process (obs) (Byers process)   An American process for making wrought iron by blowing pig iron in a Bessemer converter, pouring it into pre-melted slag and consolidating the ball so made in a hydraulic press. Recently discontinued.

Aston iron   Wrought iron made by the Aston-Byers process (qv).

Atomization   Breaking up heavy fuel oil into a spray of fine drops so that it can mix with air for combustion. It can be done by the design of the burner, which swirls the oil through a specially designed orifice; by a jet of compressed air; or by steam. The latter is the most common method.

Austenitic   An alloy iron or steel consisting mainly of austenite, a solid solution of iron carbide in iron.

Austenitic stainless   *See* Stainless steel.

Automatic gauge control (AGC)   A means of controlling the thickness of a product of a strip mill automatically. The gauge of the strip is measured continuously either as close to the exit side of the rolls as possible or, in the latest mills, in the roll gap itself, and corrections are fed automatically to the screws or hydraulic cylinders which maintain the roll gap. Extremely accurate control of gauge is possible.

Automation   A much misused word, but automation in the true sense is used extensively in the steel industry, which was one of the pioneers. It refers strictly to any process in which there is inspection of product followed by feedback of correction information as required to the process machinery. Automatic gauge control (qv) is an example of automation, which should not be confused with mechanization.

B

Backed-up mill   A four-high or cluster mill in which the work rolls are supported by heavier back-up rolls.

Backing sand   Used and reconditioned sand used to support the facing sand (qv) in a foundry mould.

Back-up rolls   Heavy rolls bearing on the work rolls to give them stiffness. Cf work rolls and four-high mill.

Back wall (1) (obs)   The rear inside wall of a square blast furnace.

Back wall (2)   The wall of an open-hearth furnace opposite to the doors.

Bad section   A fault in a rolled section, which is not true to the required shape. It can be caused by bad roll design, excessive wear of the rolls, which need dressing, or improper heating of the piece before rolling.

Bag (obs)   A leather pipe connecting the blast main to the tuyere in a cold-blast furnace.

Bag filter   A porous-walled bag not unlike that in a domestic vacuum cleaner, used for separation of fine dust from cooled furnace gases or from dust extraction systems, as in a foundry. Cf cyclone.

Balanced core   A foundry core with only one core print (qv).

Balanced steel (semi-killed steel)   A non-piping steel which is only partly deoxidized. No gas evolution is noticeable but enough gas is evolved during solidification to offset or balance the normal shrinkage.

Baldwin's oven (obs)   An old form of hot blast stove (qv).

Ball (obs) (puddler's ball)   The spongy lump of wrought iron as withdrawn from the puddling furnace, ready for shingling (qv).

Ball, drop   *See* Drop ball.

Ball furnace (obs)   Strictly a furnace for heating wrought iron scrap balls ready for rolling, but the term is often confused with mill furnace (qv).

Ball furnace pile   *See* Pile (2).

Balling (Pelletizing) (Nodulizing)   In ore preparation, forming crushed and prepared iron ore into balls by the tumbling action of a drum or disc. The balls are then sintered to form pellets ready for the blast furnace.

Balling-up (obs)   The manual forming of the iron into balls in a puddling furnace after it has come to nature (qv).

Ballistic cladding   *See* Explosive cladding.

Ball steels   Alloy steels made specially for producing the balls in ball bearings.

Ballstone (obs)   An obsolete North Staffordshire term for ironstone.

Banana-ing   A fault in steel plate rolling, when the plate emerges from the rolls curved crescent-wise as viewed from above. Cf. Rocker.

Bands (obs)   Wrought iron bands surrounding the stack of a masonry or brick blast furnace, to strengthen the structure.

Bank, bar   *See* Bar bank.

Banking   *See* Damp down.

Banking down   *See* Damp down.

Bar (1) (obs)   *See* Paddle.

Bar (2) (rod) (round)   Rolled iron (obs) or steel of circular cross section can be a bar, a rod, or a round and there is no generally accepted firm dividing line. But broadly a rod is from about $\frac{3}{16}$ to $\frac{1}{2}$ in diameter, a bar from $\frac{5}{8}$ to 3 or 4 in diameter; all larger sizes are rounds. Rods rolled in iron were sometimes called wire iron. In the iron trade the distinction between rods and bars was sometimes made by calling those rolled to gauge sizes rods and those in inch and fractional sizes bars. Rods today are usually rolled in long lengths and coiled and some sizes of bars are treated in the same way. The term bar is sometimes applied to sections which are not round in cross section but these should be called hexagons, squares, flats, etc according to shape.

Bar bank   A flat area on the mill floor where bars lie to cool after the final rolling pass. On a modern mechanized mill the bar bank itself will also be mechanized, moving the bars slowly or in short steps sideways out of the mill pass line and to the shearing and other finishing machinery. *See also* Jigging bar bank.

Bar dragger (obs) (bar drawer)   The operator who drew heated sheet bars from the furnace and took them to the rolls for rolling.

Bar drawer   *See* Bar dragger.

'Bare' (obs)   Gauge sizes for wire and rod products were often specified according to whether the article fitted tightly, loosely, etc, into the gauge. Common variants of gauges were 'bare', 'full', 'tight', 'easy', eg 'No 4 bare'. The actual interpretation was a matter for agreement between buyer and seller. In spite of its lack of precision, the usage worked well.

Barffing (Barff's process)   A chemical process for applying a rust-resisting surface to iron or steel. Steam is directed at the metal while it is red hot. It converts the surface into magnetic oxide of iron.

Barff's process   *See* Barffing.

Bar hold   *See* Tong hold.

Bar mill   A rolling mill, of any type, designed primarily to roll bars.

Barrel (1)   The main body or working part of a roll, as distinct from the necks and wobblers (qv).

Barrel (2) (cylinder)   The parallel part of a blast furnace stack, between the tapered stack proper and the bosh. Not present in all blast furnaces.

Barrel (3)   The largest part of a forging of circular cross section, usually greater in length than in diameter.

Barrelling   *See* Roll camber.

Barrow, charging   *See* Charging barrow.

Bar sorter (obs)   The man who knocked off the ends of cemented steel bars with a hand hammer ('ended' them) to examine the fracture and so to determine the degree of carburization.

Base box (obs)   *See* Basis box.

Basic Bessemer process   *See* Bessemer process.

Basic open-hearth process *See* Open-hearth process.

Basic oxygen furnace (BOF) Correctly any steel furnace in which the steelmaking chemistry is basic and oxygen is blown in for refining. It could therefore include Ajax, Kaldo, LD, Rotor, or VLN processes. In practice restricted to the LD. Term originated in the USA and now current in GB. Derivatives; basic oxygen steel (BOS) and basic oxygen process (BOP).

Basic oxygen process (BOP) *See* Basic oxygen furnace.

Basic oxygen steel (BOS) *See* Basic oxygen furnace.

Basic pig iron (basic iron) A pig iron made in the blast furnace specially for the basic open-hearth process. Is high in phosphorus (2 to 2.5%) low in sulphur (0.08%), and low in silicon (0.80%).

Basic refractories Refractories such as magnesite, dolomite, used for lining basic furnaces. Cf acid refractories, neutral refractories.

Basic slag Slag produced when high-phosphorus iron is made into steel. As it has a high phosphorus content in the form of calcium phosphate, it is ground to powder when cold and sold as an agricultural fertilizer.

Basic steel Any steel made by a process which is chemically basic.

Basin A depression in the cope of a foundry mould, connected to the sprue or downgate into which the molten metal is poured. The metal flows gently from the basin and prevents splashing or damage to the mould.

Basis box (obs) (base box) The former unit of quantity and size for production and price calculations in tinplate making. A complicated system of size classification was built up on the basis box. It was actually a unit of area, equal to 31,360 in² of tinplate. This could be 224 sheets, 14 × 10 in, or 112 sheets, 20 × 14 in, etc. The thickness was specified by weight per basis box; thus 112 sheets 20 × 14 in × 108 lb. Variations above and below this standard were indicated by the capital letters I, C, X, L, S, and D, which actually represented the 'substance' or thickness of the sheets. Now obsolete, replaced by standard area of tinplate (qv). *See* Appendix C.

Basket A large metal container with outward opening doors (petal-shaped) at the bottom, used to charge an electric arc furnace.

Bastard iron (obs)   A cheap and poor substitute for wrought iron, made of wrought iron and mild steel scrap mixed.

Bastard three-high stand   A three-high rolling mill stand in which only two rolls are being used and one is replaced by a long spindle passing right through the stand to drive the rolls in the adjacent stand.

Batch furnace   Any furnace which only deals with a batch of metal to be heated or melted and then needs recharging. All current steelmaking is done in batch furnaces. Cf continuous furnace.

Batch process   Any process which is not continuous.

Bath lancing   Introducing oxygen into the bath of a steel furnace by means of an oxygen lance for quicker refining. The lance may be hand-held and consumable (ie a simple steel tube which burns away) or mechanically retractable and water cooled as in the Ajax furnace (qv).

Bath sample (spoon sample)   A sample of steel taken with the spoon (qv) from the bath or hearth of a furnace.

Batter   See Stack batter.

BBC   See Cable iron.

BB iron etc   See Best best iron

Beak   See Anvil.

Beam roughing method   A method of rolling joists. Starting with a square bloom the rolls first make two indentations then gradually increase the lower indentation while at the same time working the flanges. Cf butterfly pass.

Beam, universal   See Universal beam.

Beans   Coke sized to about Brazil nut size and used in smiths' hearths. For lighter work such as small chain making, smaller coke known as breeze was used.

Bear (horse) (salamander)   A mass of metal found below the hearth level of a blast furnace when it is blown out. Molten metal has penetrated the hearth while the furnace was working. Hearth bricks and slag are usually mixed in with the bear.

Becking bar (expanding bar)   A heavy steel bar somewhat similar to a mandrel (qv), but used for opening out a hollow forging which is to be extended in diameter but not length. It is used with a boat tool (shaped as its name suggests) in the press.

Becking stand   A support for a becking bar (qv). Used in pairs, one each side of the press.

Bed   *See* Bowl.

Bedded-in mould   *See* Floor moulding.

Bedding   A method of mixing different ores in a blast furnace stockyard. They are laid down in layers or beds and recovered by slicing across the beds with a machine excavator.

Bedson mill   The original multi-stand continuous mill, in which the stands were arranged with the rolls alternately vertical and horizontal. The principle, dating back to 1862, is still used.

Beehive oven (obs)   An old form of coke oven. The coal charge was carbonized in a refractory chamber with a domed roof, resembling the traditional beehive. No by-products were obtained.

Behinder (obs)   In Welsh tinplate and sheet mills the back man at the roll stand; the same as catcher elsewhere.

Belgian mill   *See* Looping mill.

Bell   The bell-shaped component fitting in the cone, and used to close the top of a blast furnace.

Bell and hopper (cup and cone)   A means of closing the top of a blast furnace. The hopper is built into the furnace top and the bell hangs inside it, in close contact with the skirt of the hopper. When materials are to be charged the bell is lowered mechanically to allow the charges deposited on it to slide into the furnace. All modern furnaces have two bells and hoppers, the upper being smaller and opened more frequently, the lower being larger and only opened when a round of charges has been deposited on it. By this arrangement the furnace top is never completely open and no gas is lost.

Bell furnace (top hat furnace) (portable furnace)   An annealing furnace which can be lifted into position over a group of coils of strip. It is then sealed and heated electrically or by gas to anneal the strip. When the annealing is done the furnace is lifted off to give access to the coils.

Bellman (obs) (Scotland)   A filler at an open-topped blast furnace.

Bells   Though not now common, church and other bells have been cast in stainless and carbon steel. They were, in fact, probably the first steel castings of complex form and quite a few were made in the nineteenth century of carbon steel, crucible method. One firm, Naylor, Vickers & Co, Sheffield, is known to have made about 7,000.

Belly   The swelling in the side of a Bessemer converter, in which the molten metal collects before tapping.

Belly helve (obs)   *See* Helve.

Belt charging   Some blast furnaces are charged by a belt running beneath the coke, ore, and flux bunkers and over the tops of the furnaces – no skips are used. No British furnace is yet charged in this way but there are several in Europe.

Belt wrapper   A coiling device for cold-reduced wide strip as it leaves the finishing stand. An endless belt leads the strip and wraps it round the coiler.

Beneficiation   An imprecise term applied to any method of up-grading iron ores before smelting. If the process is carried far it is sometimes called super-beneficiation.

Bentonite   A form of clay used as a bonding material in sand moulds in a foundry.

Berlin iron   A high-phosphorus cast iron formerly used for making intricate decorative castings. There was nothing special about the iron but the Berlin founders acquired a high reputation.

Bertrand Thiel process (obs)   A steelmaking process designed to deal with irons of abnormally high phosphorus contents. It was a two-stage process, metal being tapped off to separate it from the highly phosphoric slag and then returned to the same furnace, or charged to another, for final refining.

Bessemer box   The control room for a Bessemer converter plant has been called the Bessemer box in at least one works.

Bessemer converter   The pear-shaped refractory-lined vessel in which the Bessemer process was carried out.

Bessemer iron (obs)   Pig iron made in the blast furnace specially for use in the Bessemer process.

Bessemer process   A method of making steel from molten pig iron by blowing atmospheric air through it. The oxygen in the air oxidized the carbon, silicon, and manganese in the iron. The process could be acid (for low phosphorus irons) or basic (for high phosphorus irons). Now almost extinct in Britain.

Best iron (obs)   The second grade of finished wrought iron,

Best Best iron (obs) (BB)
piled, reheated, and worked twice from puddled bar.

Best Best iron (obs) (BB)   The third or highest but one grade, three times piled and reworked.

Best Best Best iron (obs) (BBB)   Or treble best. The highest grade, four times piled and reworked.

Best Yorkshire iron (obs)   Some Yorkshire wrought irons achieved a high reputation for quality (eg Lowmoor). They were usually dry puddled from refinery iron. Other manufacturers (eg in the Black Country) marketed what they called 'Best Yorkshire' but there was nothing special about it; it was just a high-grade iron made from selected materials with special care.

Between-pass anneal   An annealing operation carried out between cold drawing or cold rolling passes.

BG   An abbreviation for Birmingham gauge (qv).

Bick   *See* Anvil.

Billet   Any rolled or forged semi-finished piece of iron (obs) or steel up to and including 5 × 5 in. Above this size the semi becomes a bloom (qv). With the advent of continuous casting billet has tended to overlap with bloom.

Billet lengths   Wood or metal spacers which lie in the hollows of a rolling mill cruciform spindle and retain the wobbler boxes in position. They are themselves retained by leather straps, wire (or even string) wound round them.

Billet shear   A mechanical shear (qv) for cutting billets.

Billy roller   A light roller carried on brackets from a hand sheet mill housing just below pass line level. It is used as a guide and rest for the piece being rolled and also for the roller's tongs.

Binder   Any substance used to bind sand for foundry moulding.

Birmingham Gauge (BG)   A gauge in which a series of numbers equals specified decimal parts of an inch. Used for sheets and hoops. Cf. British standard wire gauge. It is really obsolete, but is still quoted occasionally. *See* Appendix A.

BISRA   Formerly British Iron and Steel Research Association, now BISRA, the Corporate Laboratories of the British Steel Corporation. Gives its name to several steelmaking and working processes.

BISRA degassing process   A process developed by BISRA (qv). Molten metal is poured into a refractory-lined tundish from which it is drawn by vacuum into a refractory-lined chamber, where gases pass off leaving the steel clean. It leaves the chamber by a barometric leg and can be teemed in the normal way. Unlike other degassing processes the BISRA process is continuous.

Bite (roll bite)   The actual parts of a pair of rolls in contact with the piece and doing useful work on it. If the angle between the vertical line joining the centres of the rolls and the point of contact of the piece exceeds about 30° the piece will not normally enter the bite. This is the limiting angle of bite.

Black annealing (1) (open annealing)   Annealing in an open furnace without any protective medium. Cf. bright annealing.

Black annealing (obs) (2)   The first annealing stage in hand rolling of tinplate or sheets.

Blackband ironstone   A British (especially Scottish) iron ore in which coal is mixed with iron carbonate. The coal burnt during smelting which helped reduce fuel consumption. Now exhausted.

Black edges   Edges of strip which have become blackened by oxidation or soot during annealing.

Black forgings   Forgings which have not been machined but may have been heat treated. Cf. As forged.

Black heart   A fault in firebricks when the surface is burnt, leaving the inside partly unburnt.

Blackheart Iron   White cast iron packed in pots with a neutral packing, eg crushed slag or sand and heated to 800 to 850°C for several days. The graphite, which causes weakness in normal cast iron, is precipitated in finely dispersed form and the iron becomes ductile and strong. It is called blackheart from the black appearance of its fracture when broken. Cf. whiteheart iron, pearlitic malleable iron.

Black heat   A heat just below a dull red, which is the first visible sign of heat in iron or steel.

Blacking (1) (obs)   A paste of coke dust and water used in hearth coking.

Blacking (2) (foundry blacking)   Finely-ground coal dust or plumbago used for facing sand moulds to give a good skin to the casting.

Blacking holes   Casting defects in which the blacking sticks to, or under, the casting surface.

Blacking scab   A fault on the surface of a casting caused by the blacking flaking off during casting.

Black patch   A local patch of scale on sheet or strip, caused by incorrect pickling.

Black pickling (obs)   The first pickling of hand-rolled sheets or tinplates after hot rolling. Usually done in hot, dilute sulphuric acid but sometimes in hydrochloric.

Blackplate   Iron (obs) or steel sheet, hand rolled ready for tinplate making, after hot rolling but before pickling. Could be sold in this form instead of going on for cold rolling and tinning.

Black strip   Hot rolled iron (obs) or steel strip, as rolled, ie not pickled.

Blank (1)   A circular forging, greater in diameter than in length, for making such things as gear wheels.

Blank (2)   In powder metallurgy the 'green' pressed component before it is sintered.

Blast (wind)   The current of air supplied by an engine or blower to a furnace, eg a blast furnace or cupola.

Blast box   *See* Wind box (2).

Blast cooling (obs) (dehydrating) (dry blast)   A method used in a few places for removing the moisture from blast furnace blast. The air was chilled by mechanical refrigeration and the moisture was precipitated as frost. The air was then heated in the hot blast stoves as usual. There were several methods; the Gayley process was one.

Blast furnace   The commonest primary producer of iron from iron ore. It is a tall, refractory-lined stack-like furnace, now mechanically charged, and fired on coke. The blast is now heated to about 1,200°C but was formerly cold (ie at ambient temperature). Modern blast furnaces produce from 1,000 to 7,000 tons of iron a day.

Blast furnace gas (top gas)   The gas given off at the top of a blast furnace. Formerly burnt at the top, now collected and used to heat hot blast stoves and to fire boilers. Is a very low grade (ie low calorific value) gas.

Blast furnace metal   *See* Hot metal.

Blast furnace output index (driving rate index)   A means of measuring blast furnace performance. It is calculated from the formula: blast furnace output index,

$$\text{BOI} = \frac{Q(B + 10)}{72\,(D - 10)} \ ,$$

where Q is the daily output in tons, B is the burden weight per ton of iron in hundredweights of 112 lb, D is the hearth diameter in feet. An index of 100 equals a very good performance by present world standards.

Blast pipe (1)   On a blast furnace the pipe from the blower to the hot blast stoves and from the stoves to the bustle pipe.

Blast pipe (2)   On a cupola the pipe from the blower to the wind box.

Blazed pig iron (glazed pig iron)   High-silicon (5% or more) pig iron produced while a blast furnace is being blown in.

Bleeder (monkey)   The pipe at the top of a blast furnace through which gas can escape. The old name was monkey.

Bleeder valve   The valve on top of a bleeder (qv) which normally keeps the gas from escaping.

Bleeding ingot   An ingot withdrawn from the mould before it has solidified sufficiently; some of the liquid core runs out.

Blister steel   *See* Cementation.

Blocking (1)   Reducing the oxygen content in an open-hearth furnace bath by adding ferro-silicon or other deoxidizer to prevent further loss of carbon and avoid loss of alloying elements.

Blocking (2)   The initial rough shaping of a forging.

Blocking impression   The impression in a dropforging die which gives the metal its initial rough shape.

Block tin (obs)   The terms block and grain tin have long ago lost any specific significance, but grain tin was generally considered to be the purer. Both were used by tinplate manufacturers.

Bloodstone   *See* hematite.

Bloom (1) (obs) (blume) (salamander)   A lump of wrought iron after hammering and before rolling or otherwise working. Cf loop

Bloom (2) (obs and rare) (v)   To cog down a billet under the hammer when it was too large for the available rolls.

Bloom (3)   Strictly speaking a semi-finished piece of rolled steel between the ingot stage and the finished product, square

Bloomary (obs)

in cross section and more than 5 × 5 in. But with the advent of continuous casting blooms tend to overlap with billets (qv).

Bloomary (obs)   *See* Bloomery.

Bloomer (obs)   *See* Bloomsmith.

Bloomery (obs) (blumary) (rare) (bloomary)   A small charcoal-fired hearth for the production of wrought iron direct from the ore.

Blooming   The first operation in producing a long forging under the press. The ingot is drawn down to a long square bar or bloom and later rounded off by swage tools.

Blooming mill   *See* Cogging mill.

Bloom shear   A mechanical shear (qv) for cutting blooms.

Bloomsmith (obs) (bloomer)   The workman at a bloomery.

Blumary (obs) (rare)   *See* Bloomery.

Blow   In a Bessemer or other converter one complete cycle of operations.

Blow down   *See* Blow out.

Blow George (obs) (rare)   A small pipe temporarily connected to the blast furnace blast pipe and brought to a point where a blast of air was needed for some special purpose, eg clearing a gobbed-up furnace. *See* Gob up.

Blow gun   A hand-controlled or mechanically operated valve attached to a compressed-air line and used to blow dust, dirt, or sand particles out of a foundry mould or off a pattern.

Blowholes   Faults in the form of gas-formed cavities in or under the skin of a casting or ingot.

Blow in   The process of putting a blast furnace into commission. It is lighted with wood, the blast is turned on gently and gradually increased as the furnace burden is built up, until full blast and burden can be carried. The furnace is then 'in blast'. Cf. Blow out.

Blowing-out lines   *See* Lines.

Blowing tub   *See* Tub.

Blown ingot   A steel ingot in which the gases are released as the molten metal cools and solidifies, leaving it full of blow holes.

Blown metal   Molten metal at the end of a Bessemer blow, before the necessary elements are added to produce the required specification.

Blow out (blow down)   The process of taking a blast furnace out of commission. The burden is gradually lowered and the blast is reduced until finally it is stopped altogether and the furnace is allowed to go cold. There are several detailed procedures for blowing out. Also called blow down but 'out' is to be preferred. Cf. blow in.

Blow pipe　A pipe which conveys the blast from a blast furnace tuyere stock to the tuyere.

Blue billy (obs) (purple ore) (burnt ore)　Iron oxide residue from sulphuric acid works, used as a fettling for puddling furnaces.

Blued steel　*See* Blueing.

Blue finish　*See* Blueing.

Blue flats (obs)　A type of ironstone formerly found in the Black Country.

Blueing　Surface finishing of polished steel by oxidizing the surface lightly, when it becomes very dark blue. Can be done in various ways; eg by a flame, in hot sand, by steam.

Blume (obs)　*See* Bloom (1).

Board hammer　A drop hammer which is lifted by friction rollers gripping a wooden board fixed to the upper part of the tup. At the top of the stroke the rollers are withdrawn and the board and tup fall by gravity.

Boat guard iron (obs)　*See* Specials.

Boat tool　*See* Becking bar.

Bobbin section　*See* Specials.

Bochumer Verein process　A German method of casting an ingot in a vacuum chamber.

Body (obs)　A term formerly used to distinguish a good steel (especially in Sheffield) which was said to have 'body'. The term was never defined.

Bogie (obs)　A small two-wheeled trolley used to carry the puddled balls from the puddling furnace to the shingling hammer.

Bogie, cinder (obs)　*See* Cinder bogie.

Bogie furnace　A heating furnace in which the charge is placed on a wheeled bogie for transfer into the furnace chamber.

Boil (1) (obs)　That part of the wet puddling cycle when the carbon is burning off rapidly and the whole charge appears to be boiling.

Boil (2) (carbon boil)　That part of the steelmaking process when the carbon comes off.

Boiler plate   Iron (obs) or steel (usually mild) rolled into plates for steam boiler making.

Boilings (obs)   Slag or cinder which overflows from a puddling furnace during the boil.

Bolter-down (breaker-down)   The man who is stationed on the entry side of a two-high or three-high hand rolling mill and makes the first pass in the rolling sequence. He is usually the second in command, after the roller.

Bolting rolls   *See* Roughing mill.

Bonderizing   A proprietary method of phosphating (qv).

Bonnet (obs)   *See* Fiery steel.

Book mould   A split mould made in a hinged box which opens like a book.

Book test   A test for steel plate in which it is folded back on itself through 180°.

Booster ejector (Hogger, USA)   A high-capacity steam ejector used for a short time to help evacuate a vacuum degassing vessel quickly.

BOP   *See* Basic oxygen furnace.

Borax   A sodium-boron compound, used in powder form as a lubricant in bright drawing of steel.

Boron   A metallic element, symbol B, used in very small quantities in some alloy steels to improve their hardening properties.

BOS   *See* Basic oxygen furnace.

Bosh (1)   A water tank or trough in a forge or rolling mill, or near a furnace, used to cool the working tools.

Bosh (2) (boshes)   That part of a blast furnace which tapers outwards from the well or crucible to join the stack. It is the widest part of the furnace.

Bosh angle   The angle which the bosh of a blast furnace makes with the horizontal. It has varied greatly over the years but today about 80° could be taken as typical.

Boshes   *See* Bosh (2).

Boshing (obs)   Plunging a red-hot sheet bar or partly rolled sheet rapidly into and out of a bosh of cold water to break off scale.

Bosh rubbish (obs)   Cinder which fell from the tools and collected in the bosh. Used for fettling (qv) in the puddling furnace.

Boster (obs) (Black Country)   *See* Sheet (2).

Bott (1)   An iron or steel rod, tapered to fit the blast furnace
monkey cooler and used to stop the flow of slag from the
cinder notch.

Bott (2)   A piece of clay stuck on the end of a bar or bott stick
and used to stop the flow of metal from a cupola.

Bott stick   A steel rod with a small disc at one end to receive
the bott and a loop or handle at the other.

Bott-up (v)   To stop a cupola tap hole with a bott (qv).

Bottling   Reducing the diameter at the end of a hollow forging,
ie shaping the end like the neck of a bottle.

Bottom (1)   *See* Plug (2).

Bottom (2) (obs)   The fettled bed or bowl of a puddling furnace.
To make a bottom was to fettle it completely after a rebuild or
heavy repair, as distinct from repairing the fettling before each
heat.

Bottom blown converter   The true Bessemer converter, in
which the blast enters through the bottom. Cf. Tropenas
converter.

Bottom casting   *See* Teem.

Bottom plate (1)   The cast-iron plate on which an open-ended
ingot mould rests during teeming.

Bottom plate (2)   A loose plate in the bottom of an ingot
mould to receive the first impact of molten metal.

Bottom pouring   *See* Teem.

Bottom-pour ladle   A ladle used for ingot teeming in a
steelworks. The metal is released through a hole in the bottom,
controlled and stopped as required by a manually operated plug
called a stopper.

Bottom teeming   *See* Teem.

Bowl (obs)   The dished working bed of a puddling furnace,
which was fettled and held the charge of iron to be worked.

Bowling iron (obs)   A high grade wrought iron formerly made at
Bowling ironworks, Bradford.

**Box annealing (close annealing) (coffin annealing) (pack annealing)
(pot annealing)**   Annealing of packs of iron (obs) or steel
sheets under a heavy metal cover, resting on a sand seal to
prevent air and products of combustion from reaching them and

Box, cinder (obs)

so reducing scaling.

Box, cinder (obs)   *See* Cinder box.

Box filler (obs) (mine filler)   A worker, usually a boy or youth who filled boxes, baskets, or barrows with ore and coke for hand charging a blast furnace.

Box guide   *See* Guide mill.

Box hook (obs)   An iron hook used to tow and steer the cinder box or bogie (qv) at a puddling furnace.

Box pile (obs)   *See* Pile (1).

Bran   *See* Branning.

Brand (mark)   A maker's name or trade mark cut into the surface of a roll, so that it is rolled on to the surface of the product. In rolling iron (obs) was often called marking. Staffordshire marked bars (obs) were formerly well known.

Branded pass   A roll pass in which a brand or other mark is rolled on as the piece goes through.

Branner (branning machine)   A machine for rubbing bran on the surface of tinplate to clean it and remove oil from the surface. Other absorbent materials may be used in place of the original bran.

Branning (obs)   Rubbing the surface of tinplate manually with bran to clean it. Moss and sawdust were also used.

Branning machine   *See* Branner.

Brasses   Bearings of brass or gunmetal made in halves and used on the necks of rolling mill rolls and other rotating parts, including engines. Virtually obsolete.

Breakdown (a)   Sometimes used as synonymous with roughing (qv) but in some rolling mills a breakdown stand, performing the first rough rolling precedes the roughing stand proper.

Break down   The first stage in rolling in a bar or strip mill and, loosely, any initial rolling.

Breaker   A spherical seating on the top of the top roll chock in a rolling mill. It receives the end of the housing screw which adjusts the chock, and therefore the roll, position.

Breaker down   *See* Bolter down.

Breaking spindle   One of the spindles in a mill train which is designed to break in the event of an overload and so prevent more serious damage.

Breakout   An accidental escape of molten metal or slag from a furnace.

Breast hole (obs)   The access hole at the bottom of a solid-bottom cupola, closed when working by a wrought iron plate with the tap hole in it.

Breeze   *See* Coke breeze.

Brett's patent lifter (obs)   A steam lifting device for a drop stamp. It had a vane piston in a cylinder mounted above and connected to the tup by a strap.

Brett's patent puller (obs)   A friction drive for converting a kick-stamp (qv) to power operation.

'Brickie'   A colloquialism for the specialist bricklayer who builds and repairs furnace and ladle refractory linings.

Bridge (obs)   *See* filling place.

Bridge deck   *See* troughing.

Bridge house (obs)   *See* filling place.

Bridge rail   A heavy rolled steel rail of arch form but with flat feet, as used on bridges or for overhead crane rails.

Bridging   *See* Hanging.

Bridge stocker (obs)   A workman in charge of assembling the raw materials for a hand-charged blast furnace and hoisting them to the top or bridge for charging.

Bridle   *See* Drive bridle.

Bright annealing   Annealing carried out in a furnace with a controlled (ie non-oxidizing) atmosphere so that little or no scale is formed on the metal surface. The 'atmosphere' is an inert gas.

Bright bar   *See* Bright drawing.

Bright drawing (cold drawing)   Hot-rolled descaled bar is drawn by power through shaped dies. The bar has a bright smooth finish with close dimensional tolerances.

Bright turning   A surface-finishing operation for hot-rolled round steel bars; they are turned in a lathe or a centreless bar-turning machine.

Bring down   A term sometimes used as synonymous with melt, especially in a cupola or air furnace, where the iron is 'brought down' which means, simply, melted.

Briquette   A block of standard size made by compressing small or powdered material which is usually mixed with a binding material. Some up-graded iron ores are briquetted as are some ferro-alloys.

British Standard Wire Gauge

British Standard Wire Gauge (Imperial Standard Wire Gauge) (SWG) (ISWG) (IG)   A gauge in which a series of numbers equals specified decimal parts of an inch. Used for specifying wire diameters. This gauge is really obsolete but it is still used and is dying out only slowly. Cf. Birmingham gauge. *See* Appendix B.

Broad flange beam   *See* Universal beam.

Broadsiding (cross rolling) (length-to-width rolling)   The first few passes in rolling a sheet or plate are in a direction transverse to the longitudinal direction of the ingot or sheet bar. This spreads the piece to the required width, after which it is turned through 90° and rolled to the required length.

Brown hematite   *See* Hematite.

Brownhoist distributor   *See* Distributor.

Brushwood   Used in descaling (qv).

Bryanizing   A method of coating wire with zinc (or galvanizing it) electrically.

Brymbo desiliconization process   A method, named after the works where it was introduced, of desiliconizing blast furnace iron by means of an oxygen lance inserted into a ladle of molten iron.

Brymbo process   A method of pre-refining molten blast furnace iron with oxygen and then making it into steel in an electric-arc furnace. Devised at Brymbo steelworks, N. Wales.

Bucket charging   A mechanical method of charging a blast furnace or cupola. Instead of a skip on an incline, the charge is carried in a bucket which is lifted vertically and then traversed sideways to rest on the top or small bell. Alternatively, the bucket may travel on an incline, like a skip. The best-known arrangement for blast furnaces is the Pohlig charger.

Bucket handle section   *See* Specials.

Buckle and kink   Faults in the surface of rolled steel. Buckle is a corrugated up-and-down wrinkle; kink is a side wrinkle. Caused by faults in rolling mills or cooling beds.

Buckling   A fault in rolled bars; one side tends to be longer in the direction of rolling than the other.

Buckstave   *See* Buckstay.

Buckstay (buckstave)   A steel or cast-iron member used in furnaces to take the thrust of the brickwork.

Buggy   *See* Chariot.

Bulb angle (Butterley bulb obs)   A rolled steel angle with a bulbous end to one leg. Formerly called a Butterley bulb from the name of the Derbyshire company which first rolled it (in iron).

Bulb bar (bulb flat) (bulb plate − rarely)   A flat rolled bar with a bulbous piece on one edge.

Bulb flat   *See* Bulb bar.

Bulb plate (rare)   *See* Bulb bar.

Bulb T (Butterley beam, obs) (deck beam)   A rolled T section, the lower limb of which terminates in a bulbous piece. Formerly called a Butterley beam after the Derbyshire firm which first rolled it. Used in shipbuilding, hence the alternative name deck beam.

Bull block   A wire drawing machine comprising a power-driven capstan on which the wire is wound and a single die through which the capstan draws it.

Bulldog (obs)   Calcined puddling furnace cinder used as fettling.

Bulldozer   An American term sometimes applied to a gag press (qv).

Bull-head rail   A rolled steel railway rail with bulbous top and bottom, designed for fixing in chairs. Formerly the most common for main-line railways but now largely superseded by the flat-bottom rail (qv).

Bullet-proof plate   *See* Armour plate.

Bull's eyes   The holes in the roof of an electric arc furnace through which the three electrodes pass.

Bunker (hopper)   Both bunkers and hoppers are storage bins for bulk materials such as coal, coke, sinter, etc and there is no accepted dividing line between them. But large bins for bulk storage of raw materials are usually known as bunkers. Smaller ones are hoppers. This term is also used for bins which are purely for short-term holding of material, as in weigh-hopper, in which predetermined weights of bulk materials are gathered.

Burden   Strictly, the proportion of iron ore and flux to coke in a blast furnace charge. Sometimes loosely applied to the complete charge.

Burden chain (turning gear)   A heavy flat-link steel chain carried by an overhead crane and used to support the porter bar

Burning on (obs) (casting on)

(qv) while a forging is being made. For large forgings the burden chain is equipped with mechanical turning gear for rotating the forging.

Burning on (obs) (casting on)   A method of repairing a broken iron casting, eg the end of a roll. A mould was made of the missing part in sand and the broken end fixed on it. Molten iron was allowed to flow through the mould until the broken end melted, when the flow was stopped and the molten metal allowed to cool. A new end was thus welded on to the broken part.

Casting on was also employed for joining wrought-iron bars by means of a cast-iron boss, often ornamented, which was cast on the ends of the bars placed appropriately in the mould eg as in an iron bedstead or a traction-engine wheel.

Burning out (1)   *See* Made-up nozzle.

Burning out (2)   Cleaning gas mains of tarry deposits by deliberately setting them on fire at a time when the plant is closed.

Burnt copper   *See* Copper.

Burnt-in bottom   *See* Frit.

Burnt mine (obs)   *See* Mine.

Burnt ore (obs)   *See* Blue billy.

Burnt steel   Steel which has been overheated and permanently damaged.

Burr (flash) (fash)   Sharp, jagged edge left by a cutting tool, eg on sections hot or cold sawn.

Burst edges   Faults in cold-rolled sheet or strip in which the edges burst due to excessive rolling.

Bushelling (obs) (bustling)   Working up scraps of iron, crop ends, etc into a ball in a puddling or mill furnace for re-rolling.

Bustle pipe   The blast main which encircles the lower part of a blast furnace, and from which connexions to the tuyeres are taken.

Bustling (obs)   *See* Bushelling.

Butt   The round blunt end of a foundry rammer.

Butterfly pass   A type of pass used during the rolling of angles and channels. At one or more passes the two flanges are bent outwards into arcs; these arcs are straightened towards the end of the rolling sequence. The butterfly is the intermediate shape between the billet or bloom at the start and the finished angle or channel. Cf. flat and edging passes, notched bar passes, beam roughing method.

Butterfly valve   A valve used particularly for reversing air and gas flows in open-hearth furnaces. It consists of a pivoted disc or flap which can swing through an angle of about 90° to close two openings and open two others simultaneously.

Butterley beam (obs)   *See* Bulb T.

Butterley bulb (obs)   *See* Bulb angle.

Byers process (obs)   *See* Aston-Byers process.

By-product oven   Any modern coke oven from which the gases are taken off and the by-products extracted for commercial use.

C

Cabbage top   If an ingot is rolled or squeezed before the inside has solidified, molten metal is forced out of the end, where it finally solidifies in a rough shape known as a cabbage top.

Cabbling (obs) (US) (cubbling)   A term for cutting or breaking up pieces of wrought iron for piling or faggoting (qv).

Cable iron (obs)   Round wrought-iron bars specially rolled for making chain cables. It was often based on a regular grade eg best best which would then be called BBC (best best cable) iron.

Cage   In some very high-speed rod rolling mills the finishing stands are grouped on a common baseplate. This complete assembly is known as the finishing cage.

Calcar (obs)   A term sometimes used formerly for an annealing furnace.

Calcareous ore   Iron ore of high lime content.

Calcine (roast)   To concentrate iron ore by roasting or burning to drive off moisture, carbon dioxide, etc and to oxidize the ore to ferric oxide ready for smelting.

Calcining clamp (obs)   An open heap of iron ore and small coal, ignited and left to burn to calcine the ore.

Calcining kiln   A kiln (of various types and shapes) for calcining iron ore, usually with coal as the fuel.

Calcium silicide   An alloy of calcium, silicon, and iron used for deoxidizing steel. Usually 28 to 35% Ca, 60 to 65% Si, 6% Fe.

Calebasse (obs)   A crude form of small cupola of French origin. It had a single tuyere and no tap hole. The top of the furnace was removable, leaving the lower part, containing the molten iron, to serve as a ladle.

Calx (obs)   An old name for ore which had been calcined.

Cam barrel   *See* Helve.

Camber   *See* Roll camber.

Campaign   The period during which a furnace (particularly a blast furnace) is in continuous operation.

Cannon   *See* Stove-cleaning gun.

Cantilever mill   A rolling mill in which the working faces of the rolls are outside (or cantilevered from) the housing instead of, as in orthodox practice, between two housings. The mill pass in this case is called a cantilever pass.

Cantilever pass   *See* Cantilever mill.

Cap   *See* Capping.

Capped steel   Steel of the rimming type which has been capped. It has small blowholes below the ingot surface, in contrast to the deeper blowholes of rimming steel (qv), cast in open moulds.

Capping   Sealing the top of an ingot with a plate or cap immediately after casting a rimming steel (qv).

Capstan (1)   A multi-arm (usually four) swivelling device which collects coils of hot-rolled rod from a rod-rolling mill and swings them round as required to be taken away by mechanical transport. The coils are pushed on to the capstan arms, which will hold several coils at a time, and lifted off by crane or power truck.

Capstan (2)   The rotating drum or block which pulls wire through a drawing die.

Carbide   A chemical combination of carbon with another element (in steelmaking with iron). Metallic carbides are hard and brittle; so is iron containing them.

Carbon   A non-metallic element, symbol C, which unites freely with iron and is vital to the chemistry of iron- and steelmaking.

Carbon arc   An electrical arc passing between two carbon electrodes or between an electrode and another electrical conductor. Generates great heat and is the basis of some electric furnaces.

Carbonate ores   Iron ores containing carbonaceous materials.

Such ores are usually calcined (qv) before smelting.

Carbon boil   *See* Boil (2).

Carbon case hardening   *See* Case hardening.

Carbon dioxide (CO_2)   A colourless gas produced as a byproduct of combustion and in steelmaking. It is not poisonous but will not support life. Cf. carbon monoxide.

Carbon dioxide (CO_2) process   A method of foundry moulding. The moulds are made in special sand coated with sodium silicate and then 'gassed', ie CO_2 gas is blown through to harden the mould ready for the molten metal.

Carbon flame   The characteristic white flame of carbon burning off in a Bessemer converter. It comes after the silicon has been burnt out.

Carbon hearth   A blast furnace hearth and crucible or well made of carbon instead of clay refractories.

Carbonic acid   An old name for carbon dioxide (CO_2).

Carbonic oxide   An old name for carbon monoxide (CO).

Carbonize   To convert coal to coke by heating it in a coke oven (formerly in open heaps) to drive off gas and volatiles. Also applicable, formerly, to conversion of wood to charcoal. Should not be confused with carburize (qv).

Carbon monoxide (CO)   A colourless, odourless gas produced by incomplete combustion, and in iron and steelmaking. It is highly poisonous even in low concentrations. Cf. carbon dioxide.

Carbon steel   An alloy of iron and carbon, with less carbon than cast iron (qv). Other elements may be present in small quantities.

Up to 0.25% C is low carbon steel. From more than 0.25% C to 0.5% is medium carbon steel.

Above 0.5% and up to the upper limit, about 1.4% C, is high carbon steel.

Carburize   To introduce carbon into steel, as in case hardening (qv) or into a charge of molten steel. Under certain conditions carbon can be removed from the surface of a carbon steel. This is decarburization which has to be avoided by various means in furnaces.

Carburize should not be confused with carbonize (qv).

Carburizing flame   *See* Reducing flame.

Carburizing, selective   *See* Selective carburizing.

Cart tyre iron (obs)   The name is really self explanatory but tyre iron was rolled in various purely arbitrary sizes and designated cart, wagon, coach, or gig tyre iron according to its proposed end use.

Case   The carburized skin of a steel.
See Case hardening.

Case hardening (carbon case hardening) (surface hardening)
Introducing carbon, by one of several methods, into the skin of a soft iron or mild steel. The skin is converted into carbon steel which can then be hardened, but the core, which has not taken up carbon, remains soft. (*See* Cyaniding, Nitriding).

Case-hardening compound (CH mixture) (CH compound)
Material mainly of carbon (eg charcoal) in which steel is packed for case hardening. There are some proprietary compounds.

Case-hardening steels   Steels, which may be plain carbon or low alloy, which are specially made for case hardening after machining.

Casement section   *See* Specials.

Cast (1) (v)   To release the molten iron from a furnace by tapping it, or opening the taphole.

Cast (2) (v)   To pour molten metal into a mould.

Cast (3) (a)   Used with the name of a metal to describe any article which has been made by pouring the metal, molten, into a mould and letting it solidify by cooling, eg cast-iron flywheel, cast-steel housing.

Castable refractories   Refractory materials which can be made into a thick liquid and cast to the required shape in moulds or in situ in a furnace.

Castella beam   A proprietary (British Steel Corporation) lightweight steel beam, made by flame cutting a steel beam down the web in a saw-tooth pattern, and then welding the 'teeth' crest to crest so that the finished beam has a series of lightening holes.

Cast house   A building at the front of a blast furnace, covering the pig bed if there is one, where the casting of the iron is done.

Casting (1) (n)   An object made by pouring molten metal into a shaped mould and allowing it to solidify.

Casting (2) (v)   The whole process of making a metal casting or, alternatively, the actual pouring of the molten metal into the mould.

Casting (3) (v)   The process of tapping a blast furnace.

Casting car   A rail bound car carrying a ladle of molten steel and running over a line of ingot moulds in a pit. The ladle was of the bottom-pour type and teemed direct into the ingots. Now rare.

Casting flush   *See* Flush.

Casting on (obs)   *See* Burning on.

Casting pit (1) (moulding hole)   A pit in the foundry floor, in which large castings are made.

Casting pit (2) (pitside)   A section of a steel melting shop adjacent to the tapping side of the furnaces, in which the molten metal is teemed (cast) into ingot moulds.

Cast iron   Iron with a total carbon content of from about 1.8 to 4.5%. Also present in varying but small amounts are silicon, manganese, sulphur, and phosphorus. Used to make iron castings and as the raw material for steelmaking, wrought and malleable iron making.

Castor bed   A group of pillars 2 or 3 ft high with swivelling castors on top of each. Rolled steel plates can easily be manoeuvred on a castor bed, manually or mechanically, up to a fixed shear for shearing to size and trimming.

Cast steel (1) (obs)   A term formerly used to distinguish crucible steel (qv) from shear steel.

Cast steel (2)   An imprecise term sometimes applied to high carbon steels and also to steel castings.

Catalan forge (obs)   A form of bloomery formerly used in southern Europe. The blast was provided by a trompe, in which water falling down a pipe drew air in through a number of holes near the top of the pipe. The mixture of water and air passed into a closed chest or blast box, from which a pipe led to the tuyere. Water passed out through a hole in the bottom of the box and air went out via the tuyere pipe.

Catched ingot (obs)   When teeming crucible steel into an ingot mould the stream of molten metal had to reach the bottom without touching the sides; if it did touch, it would make a faulty or catched ingot.

Catcher   The back man at a stand of rolls in a hand rolling mill. He catches the piece as it issues from the rolls and sends it back

Cellar (1) (obs)

ready for the next pass.
See also Behinder.

Cellar (1) (obs)   The arched access space below a crucible furnace or hole.

Cellar (2) (oil cellar)   An underground room near the mill stands of a large mechanized rolling mill for housing the lubricating oil and grease tanks, pumps, and filters. It is put underground to get it as near to the mill as possible yet keep it out of the way of harm and leave the mill floor clear for working.

Cellar lad (obs) (nipper)   A boy who attended to the ashes at a crucible furnace and did the odd jobs.

Cementation (obs)   A process formerly used for carburizing soft iron bars to make a hardenable steel for knives and cutting tools. Bars were heated for several days in contact with carbon. When carburized they had blisters on the surface and were called blister steel or shear steel. Shear steel bars doubled back on themselves and fire welded were called double shear. Blister steel was also called converted steel.

Cemented steel (obs)   *See* Cementation.

Centreless grinding   Surface finishing round − usually bright drawn − steel bars by passing them between an abrasive grinding wheel and a control wheel which rotates and traverses them. Done for good finish and accurate dimensions.

Centrifugal casting (spun casting)   A casting method in which molten metal is poured into a rotating mould. It is flung to the walls of the mould by centrifugal force and solidifies there. Many cast-iron pipes and automotive cylinder liners are so made. Also called spun cast.
See also James spun cast procees. The de Lavaud process was the first commercially successful one.

CGF   *See* Controlled grain flow.

Chafery (obs)   A small hearth, fired on charcoal, coal or coke, for reheating wrought iron blooms ready for reworking.

Changing copper   *See* Copper.

Channel (channel iron) (obs) (RSC) (channel bar)   Iron or steel rolled out to a channel-shaped cross section. Iron channel is now obsolete but the term channel iron is often wrongly applied to steel channel. Rolled steel channel is often abbreviated to RSC.

Channel bar   *See* Channel.

Channel iron (obs)   *See* Channel.

Chape rods (obs)   *See* Slit rods.

Chaplet   A metal support used to hold cores in foundrywork. Chaplets should melt as the metal is poured in and become part of the casting.

Chapman producer (obs)   A gas producer based on the Siemens producer.

Chaps   The jaws of tongs (especially blacksmiths'). Cf. reins.

Char   *See* Smithy char.

Charcoal   Wood which has been distilled, leaving only carbon. Formerly used as a fuel in ironmaking.

Charcoal furnace (obs)   Any blast furnace using charcoal as its fuel.

Charcoal iron (obs)   Iron made with charcoal fuel. Could be either pig or wrought iron.

Charcoal pig   *See* Charcoal iron.

Charcoal plate (obs)   *See* Charcoal tinplate.

Charcoal tinplate (obs) (charcoal plate)   Originally tinplate made from charcoal iron, as distinct from that made from coke smelted iron; the latter was called coke tinplate or coke plate. Then the terms were applied to coke smelted plates, charcoal signifying a heavy tin coating, coke a normal coating.

Charge (1) (n)   The total quantity of iron, scrap, flux, etc put into a steel furnace during one heat.

Charge (2) (v)   The act of loading materials for smelting, melting or heating into a furnace.

Charger (1)   *See* Charging machine.

Charger (2) (obs)   A man who received the barrows of ore and coke at the top of a hand-charged blast furnace and tipped their contents into the furnace. The employment of chargers was not invariable; sometimes the filler did both hoisting and charging.

Charging barrow (obs)   A single-wheeled or latterly a two-wheeled hand barrow to carry blast furnace charges.

Charging box   A long narrow steel box designed to be picked up by a mechanical charger for placing the charge in a steelmaking furnace. It is rotated on its long axis when in the furnace to tip out the charge.

Charging door   The door in the side of a furnace (eg an open hearth) through which the materials are charged.

43

Charging hole (obs)   One of several holes in the tunnel head of an open-topped blast furnace, through which the charges were tipped.

Charging machine (charger)   Generally any machine for putting the charge into a furnace, and specifically the machine used for charging an open-hearth furnace. It may be carried by a type of overhead crane (overhead charger) or it may be run on rails or railless wheels (ground charger).

Charging plate (obs)   A piece of iron plate hung by a top hinge in the charging hole of an open-topped blast furnace tunnel head. It swung inwards when the charging barrow was tipped against it and otherwise deflected the flames up the tunnel head.

Charging skip   *See* Skip hoist.

Charging wagon (obs)   A mechanically hauled wagon running on a hand-charged blast furnace incline to carry the charging barrows to the top.

Chariot (buggy) (ingot car)   A mechanically propelled car, running on rails, to carry ingots from the soaking pits to the primary mill.

CH compound   *See* Case hardening.

Checker   *See* Spade.

Checker bricks   *See* Chequer bricks.

Checker plate   *See* Chequer plate.

Checking   Checking a blast furnace is taking the blast wholly or partly off for a short time, to lower and settle the burden or for other operational reasons.

Check nut   A nut on the housing screw of a rolling mill, which can be set to prevent the screw from moving down too far.

Cheek   A box placed between the cope and drag to make a three-part mould. There may be more than one, to make the mould a four-part, five-part, etc.

Cheese (1)   A piece cut from a steel ingot for subsequent forging. The term is most commonly used for pieces cut for making railway tyres but is also applied to pieces for other purposes.

Cheese (2)   *See* Plug (2).

Cheesing   Making cheeses (qv).

Chemically pure iron   This exists only in laboratories as a curiosity. All the commercial forms of iron have some other elements present. The nearest to a chemically pure iron in commercial form was high grade wrought iron. Today very mild steel is the nearest.

Chenot process (obs)   A long-forgotten process for direct reduction of iron ore to iron in a vertical retort.

Chequerboard roof   An open-hearth steel furnace roof formed of alternate bricks of fired magnesite and chrome-magnesite, designed to give longer life, which in practice it did.

Chequer bricks (checker bricks)   Refractory bricks shaped to form a honeycomb or open chequerwork through which gases and air can pass in a regenerator (qv).

Chequer plate (checker plate) (US)   Steel plate with a shallow pattern of diamond-shaped ribs rolled on the surface. There are several patterns, of which the Admiralty pattern (a small diamond) is the commonest. Cf. anti-slip plates.

Chest (obs) (coffin) (pot)   The clay, coffin-shaped vessel in which cementation of bar iron was carried out.

Chill (v & n)   *See* Chill cast.

Chill cast   Part of an iron casting which has been chilled or cooled quicker than the rest of the casting by means of an iron block or 'chill' moulded in the appropriate part of a sand mould. The chilled part is very hard.

Chill-cast pig (rare)   Another name for machine-cast pig iron (qv).

Chilled roll   *See* Grain roll.

Chimneying   Segregation of the coarser material in the centre of a blast furnace stack. The hot gases ascend more easily through the 'chimney' and the gas movement is sluggish round the outside.

Chimney valve   *See* Gas valve.

China clay   A natural clay found in Devon and Cornwall and used as a constituent of refractories and, formerly, of crucibles.

Chipping   Removing surface defects from billets, blooms, slabs, etc by cutting them out with a hammer and chisel or with a mechanical or pneumatic hammer.

Chisel   A tool of various sizes used by the blacksmith for cutting metal. It may be hand held or fixed in a handle, when it is called a cold set (if used for cutting cold) or a hot set (if used for hot cutting).

CH mixture *See* Case hardening.

Chock A block carrying the roll bearing and sliding in the rolling-mill housing to give vertical adjustment to the roll. Two are needed, one at each end of the roll, normally, unless the roll is cantilevered (less common).

Choke *See* Overfill.

Chop A drop forging defect caused by a piece of metal being knocked off the forging and forced into the surface.

Chromium An element, symbol Cr, used in alloy irons and steels, particularly stainless and heat-resisting steels.

Chuck *See* Porter bar.

Churning Another name for feeding (qv).

Cigar coolers Hollow iron castings arranged for internal water circulation, inserted in the stack brickwork of a blast furnace for cooling the refractory lining. Cast iron can be used at this part of the furnace, where the temperature is lower; all other blast furnace coolers are of copper.

Cigar ladle *See* Torpedo ladle.

Cinder (obs) The normal Black Country term for slag from the blast furnace, puddling and mill furnaces. Also widely used elsewhere.

Cinder bank (obs) *See* Slag bank.

Cinder bogie (obs) A small cast-iron four-wheeled box, used to catch molten cinder as it ran from the puddling furnace.

Cinder box (obs) Similar to cinder bogie and for the same purpose, but without wheels and often smaller.

Cinder fall (obs) A trough or channel leading from the blast furnace cinder notch to the roughing holes or cinder pots, ie depressions in a sand bed where the cinder was run and allowed to cool.

Cinder hill *See* Slag bank.

Cinder notch *See* Slag notch.

Cinder pig (obs) (part mine) Pig iron made from a mixture of iron ore and puddling furnace cinder, or cinder from the old direct-reduction process of wrought iron manufacture. Cf. all-mine. The term was mainly peculiar to the Black Country.

Cinder pot (obs) A shallow depression, roughly circular, near the forepart of a blast furnace for receiving molten cinder. *See* Dog (4).

Cinder snapper (rare) (snapper)   A man who cleans solid slag (cinder skulls) from slag runners.

Cinder tub (obs)   An iron truck running on rails, used for taking molten cinder from a blast furnace. It had removable sides, which were taken off when the cinder had cooled; the lump of cinder was then pushed off on to the tip.

Circle brick   A refractory brick with two sides so shaped that a number of bricks can be built into a circle, as in lining a ladle or shaft furnace.

Cire perdue process   The French name for the lost-wax or investment casting process (qv). It is sometimes found in English writing.

Cladding (sandwich rolling)   Metallurgically bonding by hot rolling two different kinds of steel, eg stainless and low carbon, to give a plate with a high quality skin and a lower grade body. Applied particularly to plate.

Clad steel   *See* Cladding.

Clamp (1)   *See* Lug.

Clamp (2)   *See* Coke hearth.

Clamp, calcining   *See* Calcining clamp.

Clamps   *See* Mould clamps.

Clay gun   *See* Mud gun.

Clay ironstones   *See* Argillaceous iron ores.

Clay wash   A mixture of clay and water used in foundries, eg for facing loam moulds.

Clean gas   *See* Raw gas.

Clean steel   Has various meanings (eg steel with an oxide-free surface) but the most common one is steel which is as free as possible from metallic inclusions, slag and other impurities.

Clear chill   *See* Grain roll.

Clear space hammer   *See* Rigby hammer.

Clink   An internal or surface crack in an ingot, caused by uneven expansion or contraction during heating or cooling.

Clinked ingot   An ingot containing clinks (qv).

Clinker   Fused or semi-fused lumps of ash and other non-combustible matter.

Clog iron (obs)   Small rolled wrought-iron sections used on the soles and heels of clogs. The term is sometimes used derisively

Close annealing

for poor quality steel; 'It is only fit for clog iron'.

Close annealing   *See* Box annealing.

Closed forepart   *See* Forepart.

Closed pass   *See* Pass.

Closed-top furnace   A blast furnace with apparatus for closing the top to take off gases. The closing device, a bell and hopper or a double bell and hopper, can be opened for charging raw materials. Cf. open top furnace.

Closed-top housing   A rolling mill housing in which the bottom, sides, and top are all in one piece.

Closer   *See* Pup.

Cloudburst hardening   *See* Shot peening.

Clough's mechanical puddler (obs)   *See* Puddling machine.

Cluster mill   A sheet or strip rolling mill which has a pair of work rolls and two or more pairs of back-up rolls. The Sendzimir mill (qv) is a form of cluster mill.

Clutch, dog   *See* Dog clutch.

CO   *See* Carbon monoxide.

CO₂   *See* Carbon dioxide.

CO₂ process   *See* Carbon dioxide process.

Coach-tyre iron (obs)   *See* Cart-tyre iron.

Coal dust   Finely ground coal used in foundries for mixing with moulding sand to form a facing sand.

Coal furnace (obs)   Coal was formerly used as the fuel for several blast furnaces in Scotland and a few elsewhere (eg North Staffs).

Coal injection   *See* Oil injection.

Coal-tar fuels (CTF)   Various mixtures of the byproducts of coking and gas making − tar, pitch, and creosote − were used as open-hearth furnace fuels but find little use now.

Cobalt   A metallic element, symbol Co, used as an alloying constituent in making steel, especially high-speed and heat-resisting steels.

Cobble (v & n)   If a piece fails to issue properly from the rolls it is said to cobble and the damaged piece is also a cobble.

Cobble shears   Mechanical shears installed in the pass line of a continuous bar or rod mill. Normally inactive, they can be brought into operation by push-button if there is a cobble at the finishing end; they cut up the material entering that end.

Although the particular billet is lost the dangers of a wild cobble at high speed are avoided and the cut-up pieces are easily handled.

Cockles (riffles)　Uneven edges on sheet metal.

Coddled iron　*See* Hot-cold.

Co-extrusion　A process for extruding iron or steel together with another metal (eg copper) to form a surface coating.

Coffin　*See* Chest.

Coffin annealing　*See* Box annealing.

Cogging　The initial working down, for subsequent rolling, of steel ingots. Usually done in a cogging mill (qv); less commonly under a steam or other power hammer.

Cogging mill (primary mill)　A rolling mill used for the initial rolling of an ingot. Confined to steelworks. Although the forge train (qv) was really a cogging mill it was never so called. The term cogging mill is falling into disuse, the primary mill being called a blooming or slabbing mill according to its use. Some roll both products and are called blooming/slabbing mills.

Coil (n & v)　When rods, bars, narrow or wide strip are made in high-speed continuous mills they emerge of great length and are almost invariably wound up into coils for convenience of handling. Bars may be cut into shorter lengths by a flying shear (qv); the others will be coiled.

Coiled plate　Some plate steel upwards of 3 mm is coiled hot and sold in this form.

Coiler　A machine which receives rolled products and makes them into coils. The commonest are upcoilers (horizontal axis, material enters from underneath) and downcoilers (similar but material comes in at the top). Bars and rods are usually coiled on a reel (qv).

Coil upender　*See* Upender

Coining (sizing)　Sizing a drop forging to close dimensional tolerances under suitable tools in a hammer.

Coke　The residue after coal has been carbonized. Used as a fuel, especially for blast furnaces and for cupolas.

Coke breeze　Small coke, ¾ in or less, screened out of blast-furnace coke, which must consist of lumps. Formerly sold to smiths but now in demand for sintering and produced by deliberate crushing.

Coke furnace   Any blast furnace using coke as its fuel.

Coke hearth (obs) (clamp)   A piece of flat open land on which blast furnace (and sometimes cupola) coke was made by controlled burning in open heaps. The product was known as hearth coke. The burning heap was sometimes called a clamp.

Coke oven   A specially designed type of oven for carbonizing coal. Is now the standard means of producing metallurgical coke. Modern coke ovens are large mechanically charged and discharged groups of coking chambers. The gas given off is a valuable fuel.

Coker (1) (obs)   A worker at a coke hearth.

Coker (2) (obs)   The member of a crucible steelmaking team who fed the furnace with coke fuel.

Coke pusher   A rail-mounted machine running alongside a coke oven battery and having a large ram, the size of an individual oven, for pushing the coke out on completion of carbonization.

Coke rate   The weight of coke consumed per ton of iron made in a blast furnace.

Coke tinplate (obs)   *See* Charcoal tinplate.

Colby furnace (obs)   An early type of electric induction furnace.

Colclad   A proprietary (British Steel Corporation) stainless steel clad steel plate.

Cold blast   Air under pressure which has not been heated.

Cold blast furnace   A blast furnace blown with air at atmospheric temperature.

Cold blast iron (obs) (cold blast pig iron)   Pig iron made in a blast furnace with unheated blast.

Cold blast pig iron   *See* Cold blast iron.

Cold blast valve   *See* Gas valve.

Cold drawing   *See* Bright drawing.

Cold finishing   An omnibus term for a variety of final operations at atmospheric temperature on steel, eg cold drawing, cold rolling, straightening.

Cold-formed steel sections   Sections formed by pressing or rolling a flat strip or sheet at room temperature to produce a component of constant cross section and of any length. Examples are sections for building, vehicles, office partitions, etc.

Cold forming   A process for making shaped components from steel. It is somewhat similar to forging and stamping, but is carried out at room temperature in a mechanical press using considerable force.

Cold-metal process   Any steelmaking process in which the complete furnace charge is of cold metal. Cf. hot metal process.

Cold-metal shop   Any steelmaking shop in which no molten metal is used in the charge, which consists entirely of cold pig and scrap.

Cold-reduced sheet (cold-rolled sheet)   Sheet which has been rolled at room temperature or cold reduced.

Cold-reduced strip (cold-rolled strip)   Strip which has been rolled at room temperature or cold reduced.

Cold reduction (cold rolling)   Rolling of steel sheet or strip at room temperature to improve its surface finish, accuracy and metallurgical properties. It is really synonymous with cold rolling but the trade nowadays prefers the terms cold reduction and cold reduced.

Cold roll corner (obs)   A fault in rolled sheet for tinplate, caused by a corner turning over and getting rolled.

Cold roll lapper (obs)   *See* Lapper.

Cold-rolled sheet   *See* Cold-reduced sheet.

Cold-rolled strip   *See* Cold-reduced strip.

Cold rolling   *See* Cold reduction.

Cold saw   *See* Hot saw.

Cold set   *See* Chisel.

Cold short   *See* Red short.

Cold shot   *See* Cold shut.

Cold shut (1) (cold shot) (shot)   A casting defect caused by non-fusion of metal where streams met during pouring.

Cold shut (2)   Freezing of the top of an ingot before teeming is complete, caused by an interruption of the stream of metal.

Collar   A projecting part of a roll barrel, used to enclose a pass.

Collaring   Sticking of a bar or other rolled section in the groove or pass of a roll. It then goes round or partly round the roll instead of issuing correctly and may cause a cobble (qv).

Collar marks   Marks on a rolled piece caused by the roll collars scoring it.

Column, universal   *See* Universal column.

Column sections   Hot rolled shapes or sections in the form of segments of a circle, with flanges for bolting together to form a circular structural column.

Combination mill   A rolling mill in which two types are combined to form the whole, eg continuous stands for the major reduction with looping stands for finishing.

Combined carbon   *See* Graphitic carbon.

Come to nature (obs)   *See* Nature, come to.

Commercial quality   Any iron or steel which is of good quality but not made to any definite mechanical or chemical specification.

Compo   An omnibus term used especially in Sheffield (it is an abbreviation of Sheffield composition) for a miscellaneous mixture of sand, fireclay, and crushed firebrick. Used for repairing refractories while they are in use; by daubing it on a thin or damaged section.

Composite iron (obs)   *See* Composite steel.

Composite roll (double-pour roll) (duplex roll)   A cast-iron roll for a rolling mill in which the outer layer and inner core are of different specifications. The outer is poured first and as soon as it has solidified to the required depth, the liquid core is run out and replaced by the second analysis, poured in molten.

Composite steel   Two steels of different analyses, eg carbon and mild, welded together in ingot form and rolled down to a finished bar or strip. Formerly common for edge-tool making — a mild steel body with a carbon steel (hardenable) working face. Also formerly composite iron/steel; wrought iron and carbon steel.

Compression   *See* Fluid compression.

Concentric converter (obs)   A form of Bessemer converter in which the mouth was at the centre of the top of the vessel instead of angled towards the side.

Condie hammer   A steam hammer with a fixed stationary piston rod and a cylinder which reciprocated and carried the tup.

Condie's tuyere (obs)   *See* Tuyere.

Conditioning   *See* Dressing (6).

Conductor rails   A special class of steel rail, with low electrical resistance, rolled specially for third-rail electrification of railways.

Constant-gap mill   A British proprietary four-high strip mill for cold rolling to very close limits. Unlike the conventional AGC (qv) the CG mill measures deviations from set roll gap in the actual gap itself as they occur, and applies corrections immediately by varying the pressure in hydraulic cylinders which set and maintain the gap.

Consumable-electrode melting   A steel-refining process in which a specially made electrode is remelted either in a vacuum or under a slag. Can be either vacuum-arc remelting (VAR) or electroslag refining (ESR) qv.

Contact angle   In a rolling mill the angle between the line joining the centres of the rolls and a radius to the top of the piece being rolled. The arc equivalent to this angle is the contact arc.

Contact arc   *See* Contact angle.

Continuous annealing (strand annealing)   Annealing of strip (wide or narrow) in a continuous furnace (qv). The nose of a new strip is joined to the tail of the old one just before it goes into the furnace, so there is a continuous ribbon of metal always in the furnace.

Continuous casting   Steel is poured into a vertical water-cooled copper mould of the shape and size of the semi-finished product required, and as it solidifies on the outside skin it is withdrawn continuously by powered rollers, further cooled by water sprays, and cut off by oxy-gas burners into mill lengths. A dummy bar is used in the bottom of the mould to start the process. Billets, blooms, slabs, and some special shapes are continuously cast and there are several types of machine. The process is less commonly applied to cast iron.

Continuous degassing   A process developed by BISRA (qv) for degassing liquid steel by sucking it continuously up a refractory-lined tube into a vacuum chamber and allowing it to run out through a barometric leg or second tube into a tundish leading to an ingot mould.

Continuous furnace   A reheating furnace in which the material to be heated is charged cold at one end and pushed or otherwise traversed mechanically through the heating zones until it emerges, properly heated, at the opposite end. Cf. batch furnace.

Continuous galvanizing

Continuous galvanizing   Galvanizing steel sheet from coils by uncoiling it, passing it through the preparation and galvanizing processes in a continuous line and recoiling it. The galvanizing can be electrolytic or hot dip.

Continuous mill (tandem mill)   Any rolling mill in which the stands are arranged in tandem and the piece goes from one stand to the next, right down the line, being reduced in each pass. Billet, bar, strip, and wide strip mills are often arranged in this way.

Continuous steelmaking   Many processes have been suggested, some have been tried but none is yet successful. Three in particular have shown promise but still have technical difficulties. The three are the British BISRA spray process, the French IRSID, and the Australian WORCRA. In the BISRA spray process a stream of molten iron falls through a ring of oxygen jets and a second ring of powdered lime jets; in the IRSID molten iron is refined inside a mass of molten slag by an oxygen lance, through which slag-making materials are also injected; in the WORCRA iron and scrap are charged into an electric arc furnace, melted and run out along a channel where refining is done by a number of oxygen jets.

Contraction rule   *See* Moulder's rule.

Controlled atmosphere (protective atmosphere)   A special atmosphere produced artificially inside a furnace to keep out oxygen from the air and so prevent oxidation or decarburization in heat treatment.

Controlled cooling   A method by which hot steel is cooled by mechanical means at a predetermined rate, as required by its metallurgical properties. It is important that some steels shall be cooled uniformly to avoid hardening, cracking, or other defects.

Controlled grain flow (CGF)   A method of making forgings so that the grain of the steel flows in the best direction to ensure maximum strength.

Conversion factors   *See* Steel conversion factors.

Converted steel (obs)   *See* Cementation.

Converter (vessel)   The vessel in which the Bessemer or other air or oxygen steelmaking process is carried out. Some countries, eg Australia, dislike the term converter as applied to oxygen steelmaking and prefer vessel.

Convex section feather edge   A rolled section, less than a half round, the edges of which are not squared; it is, in fact, a chord of a circle.

Convex section square edge   A rolled section, less than a half round, with the edges squared off. Cf. convex feather edge.

Cooler snatcher   *See* Snatcher.

Cooler, stack   *See* Stack cooler.

Coolers, cigar   *See* Cigar coolers.

Coolers, plate   *See* Plate coolers.

Coolers, stave   *See* Stave coolers.

Cooling bank   *See* Cooling bed.

Cooling bed (cooling bank)   A flat area or rack at the finishing end of a hot-rolling mill where the products are run out and allowed to cool.

Cope   *See* Moulding box.

Copper   A general colloquialism for the tuyeres and coolers at a blast furnace. 'Burnt copper' is a burnt tuyere, 'changing copper' is changing a tuyere or cooler.

Copper steels   Steels with from 0.6 to 1.5% copper content. It improves the corrosion resistance.

Core   A shape made in sand or other material, eg loam, and used in a foundry mould to form the hollow part of a casting, eg the hole in a pipe. The metal cools on the core which is then broken out.

Core bar   A metal bar built into a foundry core to stiffen and support it.

Core barrel   A skeleton former on which large or complicated cores are built.

Core blower   A machine for blowing core sand into a core box and so making a core mechanically.

Core box   A wood or metal box with a cavity of the size and shape of the required core. Sand is rammed or blown (*see* core blower) into the box. For complicated shapes the box is split into two or more parts so that the finished core can be removed.

Core drier   Sand or metal supports to keep a core in shape while it is dried.

Coremaker   A man or woman who makes cores.

Core nail   A metal nail used for supporting and reinforcing foundry cores.

Core oil   Linseed or other oil or nowadays a synthetic oil, mixed with core sand for making cores. It hardens to produce a firm core.

Core print   A projection on a foundry core which fits into a cavity of the same shape in the mould, so locating the core.

Core sand   Sand specially prepared with oil or other binder for core making.

Core stove   A special oven for baking cores.

Corrugated ingots   *See* Shapes of ingots.

Corrugated sheets   Steel sheet formed by machinery into a series of uniform ridges. Iron was formerly corrugated and the term corrugated iron is often erroneously applied to corrugated steel sheets. *See also* Pitch (1).

Cor-Ten   *See* Weathering steel.

Cort's process (obs)   *See* Puddling.

Coslettizing   A method of protecting ferrous metals against rust, by boiling in a bath of phosphoric acid and iron filings, so forming a tough iron phosphate surface coating.

Counterblow hammer (duplex hammer)   A forging hammer, worked by compressed air, steam, or hydraulics, in which the top and bottom tools are connected by steel belts, so that as one descends the other ascends to meet it; the article being forged is thus struck from above and below.

Coupler   A ring slipped over the reins or handles of tongs and pushed down tight to clamp the jaws on the work.

Coupling box   *See* Wobbler.

Cover (1) (obs)   A refractory lid placed over the mouth of a crucible furnace.

Cover (2)   The lid placed over a soaking pit.

Cover crane   A special-purpose travelling crane used for removing and replacing the covers of soaking pits.

Cowper stove   The commonest of several types of regenerative stoves used for heating the blast for blast furnaces.

Crab (obs)   *See* Dog (3).

Cramp bar (guide bar) (rest bar)   A steel bar fixed to the sides of a mill housing, parallel with the rolls, to carry the guides and guards.

Crater (obs)   Rarely applied to the throat of a blast furnace.

CRCA (cold rolled close annealed)   Steel sheets which have been cold rolled and then annealed in a closed container in a furnace, so gathering as little scale as possible.

CRCA P & O (cold rolled, close annealed, pickled and oiled) CRCA sheets which have subsequently been pickled and lightly coated with oil as a rust preventive.

Creosote *See* Coal-tar fuels.

Crocodile shear (alligator shear) A mechanical shear with two blades pivoted together so that one moves and one is stationary, ie like scissors.

Crocodile squeezer (obs) *See* Squeezer.

Cronite A cast iron for use at high temperatures; it contains nickel and chromium.

Croning process *See* Shell moulding process.

Crop ends Small pieces sheared or cropped from the ends of rolled products to square up to end, and cut off end faults.

Cropping Cutting the bad ends off an ingot, bloom, billet, or bar in a mechanical shear. Cutting into lengths in a shear is called shearing.

Crop shear *See* Shear.

Cross-country mill A rolling mill, used for bars or light sections in which the piece is never in two stands simultaneously. It issues from one pass and is skidded sideways by skew rollers and fed into the next pass in an adjacent stand.

Crosses A tinplate denomination. *See* Basis box.

Cross pile (obs) *See* Pile (1).

Cross rolling *See* Broadsiding.

Crown (crowning) An increase in thickness at the centre of a plate, sheet or strip compared with the edges, caused by the rolls bending during rolling. This is likely to be present in all such rolled products and can be a fault if too great.

Crowning *See* Crown.

Crown iron (obs) The lowest or commonest grade of wrought iron. Piled, reheated, and rolled once only from the muck bar. Often marked with a representation of a royal crown and the maker's brand, if any.

Crucible (1) (obs) (pot) A refractory vessel or pot used for melting steel by the crucible process.

Crucible (2) (well) The lowest part or well of a blast furnace, between the hearth proper and the boshes.

Crucible process (obs) (Huntsman process)   The first process for making carbon steel in which the steel was actually melted. Broken pieces of blister steel (*see* cementation) were melted in a crucible in a coke-fired furnace. The carbon mixed freely and uniformly throughout the metal, which was then teemed into an ingot mould by hand.

Crucible steel (pot steel)   Steel made by the crucible process (qv).

Crude gas   *See* Raw gas.

Crude steel   Any steel which has not been rolled. It includes ingots and continuously cast steel (which has not passed through the ingot stage). For statistical purposes the term is taking over from ingot steel, because of the growth of continuous casting. For reckoning output tonnages crude steel also includes steel for castings.

CTF   *See* Coal-tar fuels.

Cubbling (obs)   *See* Cabbling.

'Cubelow'   *See* Cupola.

Cup and cone   *See* Bell and hopper.

Cupelo   *See* Cupola.

Cupola (cupelo) ('cubelow')   A very common means of melting iron for casting. It is a shaft furnace not unlike a chimney stack, with a number of tuyeres around it near the bottom, a tap hole for releasing the molten iron, and a slag hole. A coke fire is made in the cupola and layers of pig and scrap iron are charged (usually mechanically) at the top, together with some limestone as a flux. Air blown in through the tuyeres burns the coke and melts the iron, which collects at the bottom, with molten slag floating on it. Iron is tapped off as required. Cupolas do not operate continuously, but only for a period of some hours. Most use cold air but some modern ones have heated blast. Sometimes pronounced 'cupelo' or 'cubelow'; the first is occasionally found in writing. Formerly called, at times, a 'hell', but this name is long obsolete.

Cupola block   A large refractory brick made specially for lining cupolas.

Cupola furnace (obs)   When blast furnaces were first made in the modern form, cased in iron (later steel) plates, they were sometimes called confusingly 'cupola furnaces'. The term died out, fortunately, for the furnaces so named were quite unlike a cupola (qv).

Cupolette   A very small cupola, used for small melts or for experimental work.

Cupping   A defect in wire which causes it to break with a cup-like fracture.

Cupping test (Erichsen test)   A means of testing the drawing or pressing qualities of steel sheet. A specimen is clamped between two flat rings and a round-ended ram is forced through one of the rings to push the sheet out as a dome. The height to which the dome or blister can be raised before the steel breaks or tears is the measure of its capacity to withstand pressing.

Curl end (obs)   The folded end of a pack of sheets, sheared off to separate them. The opposite side, also sheared off to square up the sheets, was the scrap or shear end. If the pack was a long one it might be sheared down the middle, forming two packs, a scrap pack and a curl pack.

Curl pack (obs)   *See* Curl end.

Curved mould   Some continuous casting machines (qv) have a curved mould and the steel issues in a large arc, to be straightened and cut off lower down. Curved moulds reduce the overall height of the machine greatly. A flexible dummy bar is necessary.

Cut-offs   A pair of knife-shaped tools used with hammer or press to cut a forging from the bar on completion.

Cutter   *See* Shearer.

Cutter-down (obs)   The leader of a team, usually three in number, cutting sheet bars to length at a cold shear for rolling to sheet.

Cutting, sand   *See* Sand cutting.

Cutting torch   A hand- or mechanically-guided torch used with oxygen and a fuel gas (eg acetylene or propane) for cutting steel.

Cyaniding   A method of case-hardening by heating steel and immersing it in a bath of molten sodium or potassium cyanide, then quenching it in oil or water.

Cyclone   A dust catcher used for cleaning blast furnace gases and for removing the dust from foundry ventilation systems. It is a cylindrical vessel with a conical bottom. Gas or dusty air is

Cylinder

introduced tangentially at the top and the dust is thrown out by
centrifugal force to the walls, whence it falls by gravity to the
cone at the bottom. It is removed periodically through an
airlock valve. Only the coarser particles can be dealt with by a
cyclone.

Cf. bag filter.

Cylinder   *See* Barrel (2).

D

Daelen mill (obs)   An early form of universal mill.

Dam (obs)   A firebrick wall at the front of an open-forepart blast furnace.

Damascene (obs) (Damascus steel)   Wrought iron and carbon steel heated together and forged into a blade by repeated heating and hammering. When polished the surface formed attractive patterns. A similar process made the Damascus gun barrels found in museums.

Damascus barrel (obs)   *See* Damascene.

Damascus steel (obs)   *See* Damascene.

Damp down (banking) (banking down)   Stopping access of air to a solid-fuel fired furnace (especially a blast furnace) to suspend its operation temporarily. In a blast furnace damping down involves taking down the blast pipes, bricking and claying the tuyeres against air ingress, and numerous other operations. It is a difficult and costly operation and there is a risk of serious derangement of furnace operation when it is put on blast again. So damping is only done when absolutely unavoidable, eg stoppage of ore supplies. A coal fired boiler or furnace is much simpler to damp down. The fire is made up heavily with fresh coal, all the dampers are closed and the job is done. Cf. slack blast.

Damper   A plate which can be let into a flue to varying degrees to control the amount of draught on a furnace. Alternatively, the damper can fit on the top of the chimney stack and be raised and lowered.

Dancer roll   A roll pivoted so that it runs on top of a strip of metal being continuously processed. It detects variations in the loop of metal formed between points in the process (eg between roll stands) and by electrical contacts and relays controls chosen variables.

Dandy (obs)   *See* Refinery.

Danks's puddling machine (obs)   *See* Puddling machine.

Darby gas offtake (obs)   A type of gas offtake used on blast furnaces. It had a large diameter tube in the centre of the furnace throat and a form of bell surrounding it. Rare.

Davies-Colby kiln (obs)   A kiln for calcining iron ore. Fired by gas, usually blast furnace gas.

Daylight   The maximum distance between the hammer and anvil or between the working faces of the tools in a forging hammer or press.

Day turn   *See* Turn.

DD steel   *See* Deep drawing steel.

Dead-blow hammer (obs)   A term formerly used, rather rarely, for a drop hammer (qv). The term is obsolete, though the hammer is not by any means.

Dead off the boil   *See* Off the boil.

Dead mild steel   *See* Dead soft steel.

Dead pass   *See* Pass.

Dead soft steel (dead mild steel)   Steel of extremely low carbon content.

Dead steel   Steel which has been fully killed. *See* Killed steel.

Decarburize (1)   To remove the carbon from cast iron as in puddling or steelmaking.

Decarburize (2)   *See* Carburize.

Deck beam   *See* Bulb T.

Deep drawing steel (DD steel)   Low-carbon steel specially rolled as sheet or strip for deep pressing or drawing; eg for motor-car body parts. Some grades are known as EDD (extra deep drawing) steel.

Definite chill roll   *See* Grain roll.

Deformed bar   *See* Indented bar.

Degas (v) (degassing)   To remove gases from liquid steel. There are several methods. *See* Continuous, Ladle-to-ladle, Stream, and Top degassing; also Dortmund-Hoerder and Ruhrstahl-Heraeus processes.

Degrease   To remove by pickling or other means the oil or grease picked up by steel during processing and so leave a clean surface for subsequent treatment such as painting, plastic coating, etc.

Dehydrating (obs)   *See* Blast cooling.

D-H process   Dortmund Hoerder process (qv).

De Lavaud process   *See* Centrifugal casting.

Delivery pipe   A steel pipe which receives the finish-rolled rod at the exit side of the last stand in a continuous rod mill and guides it to the coiling reels.

Dempster process (obs)   A process for removing tar and ammonia from blast furnace gases where coal was the fuel.

Densener   A shaped piece of iron or sometimes copper inserted into the sand of a mould to form part of the face. They are so placed as to conduct the heat more rapidly away from heavy sections, so promoting uniform cooling.

Deoxidation   *See* Killing.

Deoxidizers   Elements having a high affinity for oxygen, introduced into molten steel to take out surplus oxygen.

Derbyshire spar   *See* Fluorspar.

Descale   Any process used to remove scale from the surface of steel being hot rolled. It is often done by blowing high-pressure steam at the steel surface, or by high-pressure water sprays. Less often brushwood or heather is thrown on the surface; it explodes under rolling pressure and blows the scale off.

Deseaming   Removal of surface defects such as seams, rokes, etc from billets, blooms, slabs, etc. Done by mechanical chipping hammers or an oxy-fuel gas flame.

Descrambler (unscrambler)   A machine which takes random lots of billets or bars and feeds them forward one at a time under control as required.

Desulphurizing   Removal of sulphur from liquid iron by blowing powdered lime into it.

DH
DHHU } process   *See* Dortmund-Hoerder process.

Diagonal pass   *See* Square pass.

Diamond pass   A pass shaped like a diamond in a pair of rolling mill rolls.

Die (1)

Die (1)   A metal block with an impression cut in it, used for making drop stampings and forgings.

Die (2) (wortle plate) (wire drawer's plate)   In wire drawing a plate or block containing one or more tapering holes through which the wire is drawn. Also called, formerly, a wortle plate (obs).

Die (3)   In powder metallurgy the metal tool in which the powder is compressed to form the article ready for sintering.

Die (4)   A pierced metal plate used for extrusion.

Die block   A block of special steel used for making drop forging dies. It is itself forged, to ensure that the grain flow is correct for maximum strength.

Die-casting   Shaping molten metal by pouring it into metal moulds, which are split to allow the casting to be removed. Mainly used for non-ferrous metals but some iron is so cast.

Die line   *See* Parting line.

Die sinking   The act of cutting the impression in a die.

Die steels   A range of steels – low, medium, and high alloy – produced for making dies for press tools, plastic moulding, die-casting, and other metal-working processes. Specifications vary widely according to intended use.

Dimpling   An American term for removing a surface defect.

Dings   Kinks in the surface of a rolled sheet.

Direct casting (1) (obs)   Producing iron castings with molten iron direct from the blast furnace. The moulds may have been in the pig bed, connected by a gate to the main runner, or the iron may have been transferred by means of a ladle.

Direct casting (2)   This term is sometimes applied to the direct teeming (qv) of ingots, ie without a tundish.

Direct-fired furnace   Any furnace in which the products of combustion come into direct contact with the material being heated.

Direct process (direct reduction)   Any one of many processes in which iron or steel is produced directly from the ore, as distinct from the indirect where pig iron is first made and then purified.

Direct reduction   *See* Direct process.

Direct teeming   *See* Teeming.

Diron   A proprietary (British Steel Corporation) steel sheet specially made for one-coat surface finishing with porcelain enamel.

Dirty steel   Steel which contains impurities such as slag and non-metallic inclusions.

Disamatic   A Danish foundry moulding machine which produces sand moulds automatically without moulding boxes.

Disappearing filament pyrometer   *See* Optical pyrometer.

Discard (1) (n)   That part of the top of an ingot cropped off to remove the pipe.

Discard (2) (n)   Any unsound steel or pieces of steel rejected because of faults. It is returned for remelting.

Distributor   A mechanical device at the top of a mechanically charged blast furnace, to spread the charges evenly in the stack. There have been many types; a common one is the McKee (American). Another is the Brownhoist, also American.

Dog (1) (dogs)   The gripping jaws of a drawbench.

Dog (2)   One of the pair of jaws (dogs) which, carried on a crane, grip an ingot being charged to or withdrawn from, a soaking pit.

Dog (3) (dog clutch)   A simple clutch used in rolling mill drives. Also known as a crab (obs). It had two components (one sliding) with deep matching indentations.

Dog (4) (obs)   An iron bar, usually hooked at one end, placed vertically in the cinder pot (qv) before the cinder ran in. The cinder cooled on the dog which was used, with a crane, to lift out the solid lump for disposal.

Dog bone   *See* Preform.

Dog clutch   *See* Dog (3).

Dogger   The operator on a drawbench, who attaches the dog (*see* Dog (1)) to the bar.

Dog house   The small arched chamber at the end of an open-hearth furnace, through which the burner is inserted.

Dog leg   Strip which is bowed alternately to one side and then the other.

Dolly   Really a pair of swages (*see* Swage (1)) pivoted near a smith's anvil and normally left with the top tool swung out. When the smith wanted to complete the rounding of a piece of metal, he placed the metal in the bottom tool, swung the top one into position, and hammered it with his hand hammer. Dollies were widely used by chainmakers, for finishing or dollying a weld, but could also be found in works smithies. A works smith often had to repair chains. Cf. Tommy.

Dolomite   A form of limestone used as a refractory in basic steelmaking.

Dome brick   A refractory brick with the faces so inclined that a number of bricks can be built into a dome.

Door machine   A rail-mounted machine running alongside a coke oven battery for removing the oven doors when the coke charge is to be pushed out, and replacing them afterwards.

Dortmund-Hoerder process (DH process) (DHHU process)   A A method of degassing molten steel. A ladle is positioned below a vacuum chamber which has a refractory tube depending from it. The ladle is raised and some steel is drawn up by the vacuum. The ladle is then lowered and the steel falls back. Lifting and lowering are continued until all the steel is degassed. Instead of lifting the ladle the vacuum vessel may be lowered.

Double doubles (obs)   *See* Sheet (2).

Double-duo mill (Dowlais mill) (double-two-high mill)   A hand rolling mill containing two pairs of rolls in each stand, one above and slightly behind the other. A pass made in one pair of rolls is returned in the other. The mill functions rather like a three-high (qv) but has the advantage that more passes can be set up. It thus makes a good jobbing mill. There can be several stands, arranged in line. The term four-high is sometimes applied, incorrectly, to a double-duo mill.

Doubler (obs)   A machine squeezer for closing the fold or curl of a sheet after doubling (qv).

Double faggotted iron (obs) (double refined iron)   Terms sometimes applied to wrought iron and really equivalent to best best (qv) but imprecise and should be avoided.

Double-pour roll   *See* Composite roll.

Double-reduced tinplate (DR tinplate)   Tinplate of very thin gauges ie from 0.0066 in (0.17 mm) down. It is rolled in a continuous mill to intermediate gauge, annealed and then rolled again in another continuous mill to finished gauge. Tinning follows in the normal way.

Double refined iron (obs)   *See* Double faggotted iron.

Double refinery (obs)   A refinery (qv) used in Wales. It had a larger fire than the normal refinery and an extra set of tuyeres on each side.

Double shear steel (obs)   *See* Cementation.

Doubles (obs)   *See* Sheet (2) and Sheet sizes.

Double-sweep tin pot   *See* Tin pot.

Double-two-high mill   *See* Double duo mill.

Doubling (obs)   Part of the pack-rolling process for sheets. After a sheet had been rolled for a few passes it was doubled back on itself, reheated and rolled again.

Doubling floor (obs)   The space between the reheating furnace and the mill stand, on which the doubling was done.

Dowlais mill   *See* Double-duo mill.

Downcoiler   *See* Coiler.

Downcomer (downtake)   The pipe leading down from the offtake of a blast furnace, to bring the gas down to ground level or to the gas cleaning plant. Modern furnaces have more than one downcomer.

Down-cut shear   A mechanical shear in which the moving blade descends on the piece to be cut. The piece is often carried on a floating roller table. Cf. up-cut shears.

Downgate   *See* Gate.

Downtake   *See* Downcomer.

Down time   The time when a production unit (especially a rolling mill) is not producing, through roll changing, breakdown, or other causes.

Dowson gas   *See* Gas producer.

Dowson producer   *See* Gas producer.

Dozzle (dozzler—rare)   A refractory sleeve inserted in the top of an ingot mould, which fills with molten steel and acts as a feeder head. The dozzle is preheated before being placed in position.

Dozzler   *See* Dozzle.

D-R tinplate   *See* Double-reduced tinplate.

Draft   *See* Draught.

Drag   *See* Moulding box.

Drag-over mill   *See* Pull-over mill.

Draught (1) (draft)   The taper in the sides of a die impression to facilitate the removal of a forging.

Draught (2) (draft) In rolling, the reduction put on the piece as it passes through the rolls.

Draught (3) (draft) (pattern draw) (draw) The taper on the sides of a casting pattern to enable it to be withdrawn from the sand.

Draw (1) *See* Draught (3).

Draw (2) (draw down) The process of tapering a piece of iron or steel by smithing.

Draw (3) To pull steel bar or wire through a die and so change its cross-sectional area and possibly its shape.

Draw, pattern *See* Draught (3).

Drawn steels The act of tempering hardened steels is known as drawing the temper, and tempered steels are sometimes incorrectly called drawn steels. The term should be confined to *cold drawn* steels. *See* Bright drawn.

Drawback (false core) A body of sand in a mould which must be drawn back before the pattern can be taken out.

Drawbench A power-operated bench on which bars, rods, sections, tubes, or wire can be drawn cold.

Drawdown press A hydraulic forging press in which the hydraulic gear is in a pit below floor level and two heavy columns extend upwards through the floor to terminate in the crosshead carrying the top tool. This design is usually applied to modern oil-hydraulic presses. It ensures that if there should be an oil leak, the oil cannot spray on the hot metal being forged, so eliminating the fire risk.

Drawer The operator in charge of a drawbench.

Drawhole (obs) A term sometimes used formerly for the pipe (qv) in a steel ingot.

Drawing (1) Drawing out rods, bars, etc to wire or special shapes by pulling them by power through a succession of dies of decreasing size.

Drawing (2) (drawing down) Forging to produce a smaller cross-sectional area and therefore a greater length.

Drawing (3) Sometimes, especially in the USA, synonymous with tempering (qv).

Drawing compound Lubricants such as tallow, grease, soft soap, and oils, used in cold drawing.

Drawing down *See* Drawing (2).

Draws   *See* Shrinkage.

Dressing (1)   Sorting and cleaning iron ore.

Dressing (2)   Preparation of the surface of an ingot, bloom or slab by deseaming.

Dressing (3)   Preparation of ingot mould surfaces prior to teeming.

Dressing (4)   Cleaning the surface of a casting after fettling. (*See* Fettle 2).

Dressing (5)   Re-machining the surface of a worn rolling mill roll.

Dressing (6) (conditioning)   Removal of surface defects on semi-finished rolled steel (eg billets) by chipping, grinding, or flame gouging.

Dressing, mould   *See* Mould coating.

Drift   A blacksmith's tool somewhat similar to a punch (qv), but tapered, and used for opening out and smoothing punched holes.

Drifting   Enlarging a hole in a forging by driving a taper tool or drift through it.

Drilling (drilling up) (drilling out)   The start of the tapping operation on a modern blast furnace. A power drill is used to cut deeply into the taphole stopping, then an oxygen lance is used to hole through to the molten iron.

Drilling out   *See* Drilling.

Drilling up   *See* Drilling.

Drill, taphole   *See* Taphole drill.

Drip edge   The tail end of a tinplate which has been hot dip tinned and carries a bead of tin where it emerged from the tin pot.

Drive bridle   A group of power-driven rolls, over which a strip passes in back-and-forth movement, to impart a pull on the strip and draw it through a processing line.
A similar group of rolls but not power driven, and used solely to provide tension, is a tension bridle.

Driving   The rate of operation of a furnace, especially a blast furnace.

Driving rate index   *See* Blast furnace output index.

Drop ball (scrap ball)   A heavy cast-iron or steel ball which is lifted, usually on a tall tripod, and dropped from a height to break up scrap.

Drop-bottom cupola   A modern form of cupola in which the
bottom is made in the form of a double hinged door, held in
position while working by a heavy bar. At the end of the working
day the bar is withdrawn, the doors drop open, and the
remaining hot coke and slag fall out.

Drop forging (drop stamping)   Forming a metal shape by
repeated blows on a heated billet or bar with dies having
cavities of the required shape and size. May be done under a
power hammer or under a drop hammer, the actual hammer
part or tup of which is lifted by power and falls by gravity
between guides. Strictly drop forging should be restricted to
cases where a change of form and section of the product take
place. If only a change of form occurs the operation is drop
stamping.

Drop gate   In the foundry a pouring gate or runner which
leads directly into the top of the mould.

Drop hammer (drop stamp)   A forging machine in which the
hammer is lifted by power and falls by gravity.

Drop hammer capacity   The capacity of a drop hammer is
usually stated in terms of the weight of the tup, but the trend
today is towards indicating capacity as maximum blow energy.

Drop of the flame (flame drop)   The point in a Bessemer blow
when the intense flame ceases, indicating that the carbon is
burnt out.

Drop stamp   *See* Drop hammer.

Drop stamping   *See* Drop forging.

Drop test   A method of testing steel products, especially
railway tyres, which have to fall from a specified height (eg 5ft)
on to a specified base (eg a rail fixed to a heavy iron block)
without fracturing. Not to be confused with falling weight test
(qv).

Dry blast (obs)   *See* Blast cooling.

Dry bottom pit   A soaking pit with a floor of coke breeze
which holds the scale in semi-solid form. It is scraped out and
replaced at intervals.

Drying out   Any furnace or ladle which has been newly lined
or relined has to be dried out carefully and slowly before it
can be heated to the normal working temperature. Ladles used

to be dried by lighting coal or coke fires in them but oil or gas burners are now used. Blast furnaces can sometimes be dried out by using hot blast from an adjacent furnace, but the drying is usually done by a coke fire.

Dry plate   Defective tinplate, with dull patches on the surface.

Dry puddling (obs)   *See* Puddling.

Dry regulator (obs)   *See* Regulator.

Dry sand moulding   A foundry moulding process in which the sand, of special composition, is dried in an oven or stove, or formerly by fires built inside it, after the pattern has been removed.

Duckies (Scotland)   The pivoted fingers on a skid transfer line which 'duck' down under the rolled pieces as the transfer carriage returns.

Ductility   That property which allows metal to deform without fracture.

Dummy   An American term for a forging shaped roughly and ready for finishing between dies. Similar to the British word Use (qv).

Dummy bar   *See* Continuous casting.

Dummying (American)   Production of dummies (see dummy) for forging.

Duplex hammer   *See* Counterblow hammer.

Duplexing   *See* Duplex process.

Duplex melting   A combination of cupola and electric arc or induction furnace used for melting in foundries. Iron is melted in the cupola and tapped continuously into the electric furnace where it is held molten for use as required. The iron can also be brought to correct tapping temperature in the electric furnace and adjustments to composition can be made there.

Duplex process (duplexing)   Any steelmaking process carried out in two stages and in two types of furnace, eg blown in a Bessemer converter, transferred to and finished in, an open-hearth furnace.

Duplex roll   *See* Composite roll.

Dust bag   A cloth bag used to hold parting or blacking and for sprinkling it on a mould surface.

Dust catcher   A closed chamber in the exhaust gas system of a furnace in which the velocity of the gases falls and the heavier dust settles out at the bottom, whence it is removed at intervals.

Dwight-Lloyd process   A process for converting small pieces of iron ore into sinter (qv). The sintering is done on a travelling line (a strand) of pallets, which turn over at the end of the line to discharge the sinter.

Dye-penetrant test   A method of detecting cracks in a steel surface. More than one exists but the principle consists of painting or spraying the surface with a liquid containing a dye. This penetrates into cracks and concentrates there, so showing up.

Dynamo sheet (obs)   A term formerly used for silicon steel sheet specially made for electrical purposes.

E

Ear *See* Pin.

Easing A foundry term, meaning the removal, when the metal has solidified, of any accessible parts of the mould and/or cores, to prevent hot tearing (qv).

'Easy' (obs) *See* 'Bare'.

EDD steel *See* Deep-drawing steel.

Edge (1) (v) To turn a piece of steel being rolled through 90°, so that the next pass is made on the edges, is to 'edge' it.

Edge (2) (n) The edge of a flat sheet, plate or strip. Three conditions are recognized:

(A) Mill edge = as rolled.

(B) Sheared edge = cut after rolling.

(C) Slit edge = the edge produced in a slitting mill (qv).

Edging mill A mill stand in which the rolls are set at 90° to those in the other stands.

Edging pass Rolling the edges of a flat piece of steel either by 'edging' it as in Edge (1) or in a special edging stand, the rolls of which are at right angles to those of the other stands. *See also* Slabbing pass.

Edging rolls The rolls in an edging mill (qv).

Effervescing steel (rimming steel) A steel in which gas is evolved rapidly during the early stages of solidification in the ingot mould. The outer layers are relatively pure and the interior relatively impure. There are various types, of which rimming steel is one. Rimming steel is used for sheets which need a good surface.

Egg sleeker   A moulder's sleeker (qv) shaped like an egg.

Eights (obs)   A sheet which has been doubled (qv) three times, so forming eight sheets in pack form. Cf. fours.

Elasticity   The property of a metal which enables it to be deformed and return to its original size and shape after the deforming load has been removed. Cf. Plasticity. Steel and wrought iron can be both elastic and plastic, according to conditions.

Electrical steels   Broadly a range of sheet steels very low in carbon and having up to 4 or 5% silicon, as used in the electrical industry.

Electric arc furnace   *See* Arc furnace.

Electric furnace   Any one of several types of furnace (eg arc, resistance, or induction) in which the source of heat is electricity.

Electric pig iron   Pig iron produced from ore in an electric furnace. Made in countries where coke is unobtainable or too expensive but electricity is cheap. Not made in Britain.

Electric steel   Any steel made in an electric furnace.

Electrode (1)   A bar of carbon from which the arc is struck in an electric-arc furnace.

Electrode (2)   A specially cast or forged piece of steel or alloy used as the electrode in consumable electrode remelting or in electro-slag refining.

Electrode stub   The small piece left after an electrode has burnt away. The stub of electrode (2) is welded to a new piece of the same composition to avoid waste.

Electroflux process   *See* Electroslag refining.

Electrolytic galvanizing   Galvanizing of steel sheet with a thin layer of zinc. It is a form of electroplating.

Electrolytic pickling   Pickling in which an electric current is passed through the metal being cleaned. This induces cathodic action.

Electrolytic tinning   Coating steel with a thin layer of metallic tin. It is a form of electroplating and is the customary method of tinning tinplate.

Electrophoresis   A method of depositing metallic and other coatings on steel. It is based on the movement of electrically charged colloidal particles in a liquid when an electrostatic field is created.

Electroslag refining (or remelting) (ESR) (electroflux process) (Hopkins process)   A consumable electrode of specially prepared steel or alloy is melted by immersing the end in a shallow pool of electric-resistance-heated slag in a water-cooled mould. The electrode is lowered steadily as the end melts off and solidifies in the mould. The slag is specially prepared to refine the metal and the method of solidification is conducive to good ingot formation. Sometimes called electroflux process but the term is confusing, disliked by experts and better abandoned.

Electrostatic precipitator   A dust collector, usually the final stage in a furnace gas cleaning system. It employs an electrical phenomenon which causes fine dust particles to collect on vertical rods, tubes, or plates, from which it is removed at intervals.

Elphal process   A method of applying aluminium powder electrophoretically to steel strip and rolling and sintering it to bond it to the steel as a corrosion-resisting coating.

Elpit   A proprietary soaking pit, of Norwegian design. It is heated by electricity, the heating elements being silicon carbide troughs filled with coke.

Ending (1) (obs)   *See* Bar sorter.

Ending (2) (obs)   Breaking the end off a crucible steel ingot to see the fracture and decide the temper. *See* Temper (4).

Enfield type hammer   *See* Overhung hammer.

En specifications   A series of British Standards for steels.

Entry side (ingoing side)   That side of a rolling mill where the piece makes its first entering pass. It finally emerges from the opposite or exit side (outgoing side).

Entry table   A flat area, usually with live rollers on the entry side of a rolling mill, for feeding the pieces to the rolls. Cf. run-out table. Cf. approach table.

Epoxy resins   A plastics material now used extensively for foundry patterns and coreboxes, especially with automatic moulding machines. The material is used both in conjunction with and in place of, metal patterns, ie some patterns are a composite of both materials. It is customary to make a master epoxy resin pattern, from which working patterns can be reproduced as often as necessary.

Equal angle   *See* Angle iron.

Equal-leg angle   *See* Angle iron.

Erichsen test   *See* Cupping test.

ESR process   *See* Electroslag refining process.

Etchant   A chemical reagent used for etching (qv).

Etching   Revealing the structure of metals by selective chemical attack of the surface. If steel is polished and etched it will be found to have light and dark portions, the dark ones containing the carbon. The darker parts are also the harder ones.

European malleable iron   A term sometimes used for whiteheart iron (qv).

Exact lengths   There is no such thing as an exact length, though the term was and is often used in contracts for rolled products. In the past 'exact' was a matter for agreement between maker and buyer. Today a tolerance is usually specified in the appropriate British Standard. Cf. Rolling margin.

Exit side   *See* Entry side.

Exit table   *See* Run-out table.

Exothermic   A chemical reaction in which heat is evolved. The oxygen steelmaking processes are exothermic.

Expanding bar   *See* Becking bar.

Explosion doors   Counterweighted doors fitted at the top of a blast furnace to open in the event of an internal explosion, relieve the pressure and prevent damage.

Explosive cladding (ballistic cladding)   A method of joining plates for cladding (qv). The two plates are laid together and an explosive is placed on the top. When the explosive is detonated the explosion wave passing over the surface bonds the two metals together and they can be rolled as one to finished size. The process is new and not yet widely used.

Explosive tapping (jet tapping)   Tapping an open-hearth furnace by placing an explosive charge against the taphole and detonating it by remote control. Not very widely practised.

Extra-deep-drawing steel   *See* Deep-drawing steel.

Extrusion (1) (v)   Forcing metal through a specially shaped hole in a die so that it emerges with a cross section the same as the hole. Steel is extruded both hot and cold and needs heavy pressure. Special shapes and tubes in more expensive alloys are extruded.

Extrusion (2) (n)   Any product made by the extrusion process.
Eye of the furnace   *See* Tuyere cap.
Eyesight elbow   A small projection at the back of the gooseneck (qv) containing a blue glass window for viewing the furnace through the tuyere.

F

Fabric bearing   A plastic-impregnated fabric shaped to the necessary curvature for a rolling mill bearing and used in pairs on the roll necks.

Facing sand   Any specially prepared sand which is placed next to the pattern in making a mould. The rest of the mould is made up with used or lower quality sand, called backing sand.

Faggot (1) (obs) (fagot)   A bundle of small wrought iron bars heated and welded under a power hammer to make a shaft or other forging.

Faggot (2) (obs) (fagot)   A bundle of bars of blister steel (qv) ready for heating and welding under the power hammer to make shear steel.

Faggoted iron (obs) (fagoted iron)   Wrought iron bars or forgings made by faggoting. *See* faggot (1).

Faggoting (obs) (fagoting)   Working bundles of wrought iron or blister steel under the power hammer. *See* faggot (1) and (2).

Fagot (obs)   *See* Faggot (1) and (2).

Fagoted iron (obs)   *See* Faggoted iron.

Fagoting (obs)   *See* Faggoting.

False core   *See* Drawback.

False piece   *See* Loose piece.

Falling slag   A special slag used as a finishing slag in basic electric arc steel melting. On cooling it falls into a powder.

Falling-weight test   A means of testing steel articles, especially railway tyres. A specified weight is allowed to fall on the tyre from a specified height. There must be no fracture and there is

Fancy iron (obs)

often a specified deflexion. Not to be confused with drop test (qv).

Fancy iron (obs)   Correctly applied to rolled iron sections with indentations or patterns on the surface for making fancy ironwork (eg garden seats, fenders) but sometimes loosely used for such things as horseshoe iron.

Farnley iron (obs)   A very high grade wrought iron formerly made at Farnley Ironworks, Leeds.

Fascold   A British proprietary process and machine for making foundry cores. It mixes sand, liquid resin, and catalyst automatically and blows them into a core box of the required size and shape. The treated sand hardens quickly by chemical reaction at ambient temperature and the finished core can be taken out in a short time – from 15 seconds to 2 minutes according to the type of resin.

Fash   *See* Burr.

Fat sand   Foundry moulding sand containing a large amount of clay or alumina.

Feather edge section   *See* Convex section, feather edge.

Feeder head (feeding head)   A large riser (qv) on a foundry mould to act as a reservoir of molten metal to feed the casting as it shrinks while cooling, so preventing voids in the solid metal.

Feeder rod   A steel bar pushed manually in the feeder head to promote flow of hot metal into the mould and to break up pockets of gas.

Feeding   Supplying molten metal to a casting or ingot to fill the cavities which would otherwise form as the metal cooled.

Feeding head   *See* Feeder head.

Feed Core   High-grade, high-iron-content ore, used in small lumps or pellets added to the furnace bath as a decarburizer in steelmaking.

Feedstock   This term, common in chemical engineering, is being borrowed by the iron makers, and the charge materials of a blast furnace (iron ore, limestone, coke, and air) are sometimes referred to as the furnace feedstock. It is rather unnecessary and should be discouraged.

Fence iron (obs)   A rather low grade of wrought iron rolled into bars and sections such as stars or small channels for making fences and hurdles.

Ferrix oxide   One of the forms of iron oxide, Fe_2O_3. Cf. ferrous oxide and magnetic oxide.

Ferritic stainless   *See* Stainless steel.

Ferro   Used as a prefix with the name of an element (eg ferro manganese) for a range of alloys used for adding other elements to steel.
The most common ferro alloys are listed alphabetically by elements with the prefix ferro.

Ferro alloys   *See* Ferro.

Ferro boron   An alloy of iron and 10 to 25% boron, used to introduce boron into malleable iron and steel.

Ferro chromium   An alloy of iron and 60 to 75% chromium with up to 1% carbon. Used in making stainless and other high-chromium steels.

Ferro manganese   The usual alloy for adding manganese to steel. It contains 70 to 80% manganese and about 7% carbon, the rest is iron.
A variety known as spiegeleisen contains 12 to 30% manganese and about 5% carbon.

Ferro molybdenum   An iron alloy with 50 to 60% molybdenum. Used for adding molybdenum to steel.

Ferro silicon   An alloy of iron with from 15 to 95% silicon. Used for adding silicon to steel.

Ferrostan process   A method of producing tinplate electrolytically. Cold-reduced wide strip passes continuously through the plating bath. It is re-coiled and cut into sheets later.

Ferrosteel   *See* Semi-steel.

Ferrous oxide   One of the forms of iron oxide, FeO. Cf. ferric oxide and magnetic oxide.

Ferro vanadium   An alloy of iron with from 35 to 55% vanadium, 1.5 to 2% silicon and 0.2 to 3.5% carbon. Used for adding vanadium to steel.

Fettle (1)   To put the reactive oxide material into a puddling or steel furnace.

Fettle (2)   To cut or break off the runners, risers, and gates from a casting after it has cooled and been knocked out of the mould. Cf. dressing (4).

Fettle (3)   Used loosely to mean to put something in order or to repair it.

Fettling (1) (n)

Fettling (1) (n)   A generic term for the materials used to fettle a furnace. *See* Fettle (1).

Fettling (2) (v)   The act of putting the fettling materials in a furnace.

Fettling machine   A machine which can be lifted by the shop crane to the door of an open-hearth furnace. It has a hopper which holds fettling material and electric drive which throws the fettling through the door on to the hearth.

Fiery steel (obs)   A term used in crucible steelmaking for steel which was over oxidized and spat out sparks in the crucible. It would boil up in the ingot mould and form on the top of the ingot a worthless lump of spongy metal (a bonnet).

Fifteen carbon   Written '15 carbon'. A common colloquialism for 0.15% carbon steel. Similarly 20, 25, 30, 40, 50 and 55 carbon for 0.2, 0.25, 0.30, 0.40, 0.50, and 0.55% carbon steel. *See also* Carbon steel.

Fifty carbon   *See* Fifteen carbon.

Fifty-five carbon   *See* Fifteen carbon.

Filler (obs)   A workman who wheeled the barrows of raw materials and tipped them into a hand-charged blast furnace.

Fillet   The rounded corner of a square pass cut in a rolling mill roll.

Filling engine (obs)   The engine which wound the charging wagons up the incline on a hand-charged blast furnace.

Filling hole (obs)   The aperture in the tunnel head of a hand-charged blast furnace through which the charges were tipped. Small furnaces had one; larger furnaces several.

Filling place (obs)   The platform at the top of a hand-charged blast furnace giving access to the tunnel head for charging. Also called the bridge, but this should be confined to the platform joining two or more furnaces, Sometimes covered by a building (the bridge house) on small hand-charged furnaces.

Fine metal (obs)   *See* Refined iron.

Finer (obs)   A worker at a finery (qv).

Finer's metal (obs)   *See* Refined iron.

Finery (obs)   A hearth used to make wrought iron by the direct reduction process (ie direct from the ore). It was charcoal fired and an air blast was used to decarburize the pig iron. Cf. refinery.

Fines   A term of variable meaning but especially referring to the small pieces screened out of iron ore, sinter, or coke.

Fingers   Pivoted short arms on a mechanical transfer bank, which pass under the piece in one direction and drag it sideways in the other direction.

Fingers, tilting   *See* Tilting fingers.

Finished iron (obs)   Wrought iron rolled out into bars or sections.

Finished ironworks (obs)   Any ironworks which made finished iron (eg bars, rods, sections, sheets, plates, etc) as distinct from one which produced only pig iron or wrought iron blooms.

Finished steel   Any steel product which is ready for the market without further processing.

Finishing cage   *See* Cage.

Finishing mill   Any rolling mill other than a forge train or primary or intermediate mill; ie it produces finished products as distinct from semis (qv).

Finishings   Final alloy additions to the bath in a steel furnace.

Fire (obs)   Each charge of coke to a crucible furnace was called a fire. It needed three or four for a complete melt. *See also* Killing fire.

Firebox (obs)   *See* Fireplace.

Firebrick   A generic term for shaped bricks made from refractory clays.

Fire bridge (obs)   The wall or division between the fuel and the metal in a reverberatory furnace. Cf. flue bridge.

Fireclay   Any natural clay having a fusing point of more than about 1,600°C. Used for making refractory bricks. A typical composition is $Al_2O_3.2SiO_2.2H_2O$ (alumina, silica, and water).

Fire hook (1) (obs)   A hook-ended bar used by a puddler to level and trim his fire. Sometimes used for fettling as well.

Fire hook (2)   A form of poker used by a blacksmith to open and clean the hearth fire.

Fireplace (obs) (firebox)   That part of a solid-fuel-fired furnace (eg a puddling or mill furnace) containing the burning fuel.

Fire stone (obs)   A natural refractory rock formerly cut into shapes and used for furnace lining, especially the hearths of blast furnaces.

Fire waste   *See* Heat waste.

Fire welding   *See* Forge welding.

Firing hole   A hole in the side of a coal-fired furnace giving access to the fire.

First-hand melter   The man in charge of the team at a steel furnace. His assistants are the second- and third-hand melters.

Fishbellied rails (obs)   Railway rails which curved out on the underside from each end, so giving a 'fish-bellied' appearance in side elevation.

Fishplates   The steel sections rolled specially for making railway fishplates are themselves so called, though they are made in long lengths and do not really become fishplates until cut up to the shorter lengths required by railway engineers.

Fishtail   If a piece of steel is rolled when the skin is appreciably hotter than the centre, a V-shaped notch will form at the end. This is a fishtail and the piece is said to have fishtailed.

Five-part system   *See* Hot pack rolling.

Flaking   *See* Spalling (1).

Flame   The flame from a burning fuel can be oxidizing, reducing, or neutral. If there is more air than is needed for combustion, oxygen from the excess air will cause oxidation or scaling of the metal being heated. If there is a deficiency of air the flame will be smoky or reducing and will tend to de-oxidize the metal being heated. If there is just sufficient air for combustion the flame is neither oxidizing nor reducing, but neutral.

Flame cutting   Cutting steel with the flame from an oxy-fuel gas torch. It is used also in general engineering, but is widely applied in steelworks.

Flame drop   *See* Drop of the flame.

Flame gouging   A modification of flame-cutting, used for cutting surface faults out of billets, blooms, and slabs before re-rolling.

Flame scarfing   Practically the same as flame gouging but may be mechanized.

Flash (1) (v) (obs)   To fire a puddling furnace lightly with small coal and fully-open damper to give a sudden increase of heat.

Flash (2) (n)   A rib of iron or steel round a casting or forging where the metal has run molten or squeezed hot between the parts of the mould or of the dies.

Flash (3) (v)   To remove the flash by grinding, or chipping, or mechanically.

Flash (4)   *See* Overfill.

Flash (5)   *See* Burr.

Flash line   The line left on the side of a forging after removal of the flash.

Flash melting   *See* Flow brightening.

Flask   Another name for a moulding box (qv).

Flat (n)   Any piece of rolled iron (obs) or steel of small size and rectangular cross section, rather thick in relation to its width. There is no precise definition.

Flat and edging passes   A sequence of passes for rolling angles. The piece is progressively reduced but is kept flat until the final two passes when the edges are turned to form a right angle. Cf. butterfly pass, notched bar method.

Flat-bottom rail   A rolled steel railway rail with a bulbous head and a flat foot, for fixing either direct to a sleeper or on a small flat pad. Now the most common type for main-line railways. Cf. Bull head rail.

Flat-rolled products   Plate, sheet, and strip, the last two being either hot rolled or cold reduced.

Flat sheet (flattened sheet)   Steel sheet flattened to a special degree usually by stretching. Iron sheets used to be flattened in the same way, one at a time in a hydraulic or mechanical stretching machine; they were called patent flattened (obs).

Flattened sheet   *See* Flat sheet.

Flatter   A form of set (qv) with a flat working face, used by the blacksmith for drawing down and for removing surface inequalities.

Flexible dummy bar   *See* Curved mould.

Floor moulding (bedded-in mould)   When the mould or at least the major part of it is made in a sand floor instead of a moulding box. A box may be used on the top to contain the runner and to ensure that the top surface of the casting is correct.

Flop forging   An American term for forging in dies which are identical top and bottom, and the forging can thus be turned over between blows.

Flourishing (obs)   A process used in the period of transition

Flow brightening (flash melting)
from finery/chafery working to dry puddling, in which desiliconizing was due before decarburization.

Flow brightening (flash melting)   Imparting a bright finish to electrolytic tinplate by momentarily melting the tin coating, usually electrically.

Flue   A passageway, lined with or made of refractory material, between a furnace and its chimney stack.

Flue bridge (obs)   A refractory wall between the metal and the flue leading to the chimney in a reverberatory furnace. Cf. fire bridge.

Flue cinder (obs)   Cinder (qv) which has overflowed and been tapped from the flue of a reheating furnace, especially a mill furnace.

Fluid compression   A method of improving the soundness of an ingot by applying heavy pressure to the top surface while the steel was still molten. There were many methods, Whitworth's, using a hydraulic press, being the best known.

Fluidized bed   A bed of finely-divided solids kept in constant movement by a gas blown in from underneath, so behaving like a liquid. Fluidized beds are used as the heating device in some heat treatment furnaces, the pieces to be heated passing into or through the 'fluid' solids.

Fluorescent crack detection   A non-destructive test. A penetrating oil containing fluorescent material is applied to the surface, the surplus is washed off, and any remaining in cracks glows when viewed under black light, ie light just beyond the violet of the visible spectrum and therefore normally invisible to the naked eye.

Fluorspar (spar)   Calcium fluoride, CaF_2, often used as a steelmaking flux in conjunction with limestone. A well-known British source is Derbyshire; hence the name Derbyshire spar.

Flush (1)   To slag a blast furnace. The last flush before casting is the casting flush.

Flush (2)   In America a method of slag removal from an open-hearth furnace.

Flushing hole   *See* Slag notch.

Fluted ingots   *See* Shapes of ingots.

Flux (1)   Material, eg limestone, used in the blast furnace to combine with the impurities and form a liquid slag which can be tapped off.

Flux (2)   Material used similarly in steel furnaces. Examples, lime, limestone, fluorspar.

Flux (3)   Zinc chloride, with or without an addition of ammonium chloride, floating on the molten tin bath in hot-dip tinning to clean and prepare the sheet for the tin coating.

Flux stick (obs)   A steel or iron bar held against the mouth of a crucible to keep back slag while the metal is teemed.

Flying micrometer (flying mike)   An instrument in contact with and continuously measuring the thickness of a wide or narrow strip as it emerges from the rolls. It can indicate thickness on a dial or record it on a graph or do both.

Flying mike   *See* Flying micrometer.

Flying shear   Any form of shear which cuts rolled steel as it is still moving. The flying shear travels with the piece, cuts it and then goes back to its starting point at high speed ready to make the next cut.

Foamed slag   Blast furnace slag made into a type of foam by blowing steam through it while it is molten. It is used for thermal and acoustic insulation.

Foaming slag (sponging slag)   Under certain conditions the slag in a steel furnace will froth up to a foam, which can overflow from the furnace and be very troublesome.

Fog quenching   Quenching (qv) steel in a fog or mist of vapour instead of in a liquid.

Ford & Moncur stove (obs)   A type of blast furnace stove, regenerative but arranged internally for ease of cleaning.

Forge (1) (obs)   Buildings and machinery used for the conversion of pig iron into wrought iron. It comprised the furnaces, hammer, and forge train but not the finishing mills.

Forge (2)   Buildings and machinery used for producing iron (obs) and steel forgings. If the forgings were small or medium sized, the buildings etc are better called a smithy.

Forge (3)   This term is sometimes, wrongly, used for a small smith's shop, which should be called a smithy.

Forge (4) (v)   To shape iron or steel by hammering it or squeezing it in a press.

Forehammer (Scot)   A name sometimes given to a sledge hammer.

Forehearth (1)

Forehearth (1)   *See* Receiver.

Forehearth (2) (obs)   On the open-forepart blast furnace the opening above the dam up to about tuyere level was called the forehearth as distinct from the whole front of the furnace which was the forepart.

Forepart   The lower front of a blast furnace and often the area immediately in front of it. May be closed (today) or open (formerly) (qv). The closed forepart was sometimes called a Lührmann or front, after the inventor but this term is now obsolete.

Foreplate (1) (obs)   The front horizontal plate at the working door of a puddling furnace.

Foreplate (2) (obs)   Another name for tymp (qv).

Forge hammer   Any hammer (tilt, helve, or steam) used for shingling puddled balls. Not to be confused with forging hammer (qv).

Forge iron (obs)   *See* Forge pig.

Forge pig (obs) (forge iron)   Pig iron made specifically for the forge, for decarburizing into wrought iron.

Forge rolls (obs)   *See* Forge train.

Forge train (obs) (forge rolls) (puddled ball rolls)   The mill used in a forge for rolling puddled blooms to muck bar.

Forge welding (smith welding) (fire welding) (hammer welding)   The oldest method of welding, by heating two pieces of wrought iron or mild steel to welding heat and forging them together under the hand or power hammer. The smith's method.

Forging hammer   Any hammer (tilt, helve, steam, pneumatic) used for forging as distinct from shingling. Not to be confused with forge hammer (qv).

Forging ingot   An ingot made specially, often of distinctive shape, for forging.

Forging manipulator   *See* Manipulator (1).

Forging press   A large hydraulic (usually) or mechanical press used for making heavier forgings.

Forging rolls   *See* Gap mill.

Fork   A two-pronged hand tool used in hand rolling mills for turning bars between passes.

Forty carbon   *See* Fifteen carbon.

FOS process (fuel-oxygen-scrap process)   A steelmaking process using selected scrap as the raw material, melting down with a mixture of oil and oxygen and refining with oxygen.

Foundry   The complete plant used for the production of iron or steel castings. Sometimes misused for any shop where metal is melted. Its proper meaning is as defined in the first sentence.

Foundry blacking   *See* Blacking (2).

Foundry iron   *See* Foundry pig.

Foundry pig (foundry iron)   Pig iron made specifically for remelting in foundries.

Foundry returns   *See* Arisings.

Four-high mill (1)   A rolling mill with four rolls in each stand mounted one above the other in a vertical line. The two rolls at the pass line (the work rolls) do the actual rolling and the two outer (back-up rolls or back-ups) are much heavier and provide backing support for the work rolls. Used in wide and narrow strip mills, hot and cold. *See also* Three-high; Two-high.

Four-high mill (2)   *See* Double-duo mill.

Four-part system   *See* Hot-pack rolling.

Four-post furnace   *See* Free-standing furnace.

Fours (obs)   A sheet which has been doubled (qv) twice, so forming four sheets in pack form. Cf. eights.

Fracture   The surface of metal when it is broken by force rather than cut. The appearance of the fracture of iron and steel was formerly used extensively as a means of determining and grading the quality. Cf. number one, etc, iron.

Fracture grading   *See* Number one.

Freckles   Dull spots on tinplate.

Free carbon   Carbon which is not chemically combined with iron and appears as graphite.

Free-cutting steels (free-machining steels)   Steels to which certain elements, such as sulphur, lead, selenium, or bismuth have been added to promote machinability.

Free-machining steels   *See* Free-cutting steels.

Free-standing furnace   A blast furnace in which the whole top structure is carried on independent columns, the furnace shell having little or no load on it. Widely used in Europe but rare in Britain. A four-post furnace is an example, the four posts carrying all the top equipment.

Freezing point   The temperature at which a fluid (in this case

molten iron or steel) solidifies. It is the same as that at which
the solid melts, ie the melting point.

Friction drop stamp   A drop stamp operated by a friction
lifter (qv).

Friction lifter   A clutch fixed over a drop stamp and arranged
to be operated by a hand-pulled cord. When the clutch is
engaged it turns a shaft and lifts the tup by means of a strap.

Frit (v)   To melt and consolidate as a whole the hearth bottom
of a furnace made of paste or powder refractory by heating it in
the furnace while it 'frits'. A bottom so made is a fritted
bottom or a burnt-in bottom.

Fritted bottom (burnt-in bottom)   *See* Frit.

Fritting   The bonding together of materials by subjecting to
heat. It applies particularly to refractories used in furnace
linings.

Frontal helve (obs)   *See* Helve.

Front wall   The wall of an open-hearth furnace which contains
the doors.

Fuel-oxygen-scrap process   *See* FOS process.

Fulbond   A proprietary form of fuller's earth used for bonding
moulding sand.

'Full' (obs)   *See* 'Bare'.

Fuller   A two-part tool for making grooves in forging or for
drawing down (qv). One is held in the anvil, the other is hand
held and struck with the hammer.

Fullered horseshoe iron   *See* Horseshoe iron.

Fullering   Using the fullers (qv) for grooving or drawing down.
See Fuller.

Full-finishing blackplate (uncoated tinplate base)   Steel sheet
which has had the full processing usually given to tinplate but
has not been tinned.

Full strip   Strip which is buckled in the centre because, due to
faulty rolling, the edges are shorter than the middle. Cf.
twisted strip.

Furane resin   A chemical bonding agent used in foundry sand.
Sand treated with furane is moulded in the normal way and sets
hard enough for casting just by exposure to the air.

Furnace   Any specially designed and constructed device for
heating and/or melting iron or steel. It is always provided with

an external source of heat, eg coal, oil, gas, electricity. A Bessemer or oxygen converter, in which the heat is obtained from an exothermic reaction, is not a furnace, though it is often so called.

Furnace bank (high line)   An area at the back of a blast furnace plant, usually with a railway track or tracks, where raw materials can be assembled for charging. It may be formed from the natural lie of the land, or artificially. If the terrain is flat the rail track may be carried over a line of bunkers, with a wagon lift at one end and a lowering arrangement or wagon drop at the other. Such an arrangement is often called a high line, especially in America.

Furnace cooling   Slow cooling of steel after heat treatment by withdrawing the source of heat and allowing the furnace and its contents to cool naturally.

Furnace keeper   *See* Keeper.

Furnace lining   Refractory material on the inside of a furnace to resist the action of heat and chemical and erosive attacks by the contents. There are many forms according to the particular conditions and they are all renewable.

Fuse   Strictly, to melt.

Fusible cones   *See* Seger cones.

Fusion   The change from solid to liquid state, when the metal reaches its melting point.

G

G A general abbreviation for gauge (qv). It is not always certain which gauge is meant; the term was, and still is to some extent, used very loosely.

Gag (obs) A prop used to support a helve or tilt hammer clear of the cams when it was not in use. Inserting the gag was called gagging up. If a forging was too large for the space available under the hammer, it would have insufficient fall to be effective and the hammer was then said to be gagged.

Gagged up (obs) *See* Gag.

Gagger (iron) (lifter) An iron or steel bar, usually bent or hooked, built into a sand mould to strengthen it.

Gagging up (obs) *See* Gag.

Gag press A mechanical or hydraulic press used for straightening rolled steel bars or sections. Sometimes called in America a bulldozer.

Galvanize To coat iron or steel with zinc to protect it from corrosion.

Galvanized iron (obs) Large quantities of iron sheet were formerly galvanized, and usually corrugated, for roofing and other building purposes. It is now obsolete, having been replaced by galvanized steel sheet, but the old term lingers on, often being applied, incorrectly, to the present steel product.

Galvanizing bath *See* Galvanizing pot.

Galvanizing pot (galvanizing bath) The metal vessel containing molten zinc for galvanizing.

Galvaprime A proprietary (British Steel Corporation)

Galvatite
galvanized corrugated steel sheet, factory painted on one side and ready for use without in-situ painting.

Galvatite A proprietary name (British Steel Corporation) for continuously hot-dip galvanized steel sheet.

Gamma ray *See* X-ray.

Gangue (vein stuff) Waste material in iron ore. Term borrowed from the metal mining industry.

Ganister (gannister) A refractory highly siliceous rock used, when ground and mixed with water and fireclay, as a refractory paste. Also made into firebricks.

Gannister *See* Ganister.

Gantryman (obs) A workman on the rail gantry spanning a line of calcining kilns. He was responsible for charging the kiln with ore and small coal.

Gap mill (roll-forging machine) (use rolls) (forging rolls) A machine for making semi-finished forgings or uses. Forging is done between rolls with a series of impressions into which the work is inserted in sequence.

Garrett mill (obs) A rod mill for producing rod direct from billets. It combined a three-high billet mill and trains of finishing rolls in echelon.

Gartsherrie process (obs) (Alexander & M'Cosh process) A process for recovering tar and ammonia from a blast furnace using raw coal instead of coke.

Gas barrel (obs) A former term for tubes.

Gas carburizing Introduction of carbon into the outer layers of mild steel by heating it in a furnace, the atmosphere of which is rich in carbon compounds, such as carbon monoxide or hydrocarbon gases.

Gas cleaning Passing blast furnace, steel furnace, or other gases through a variety of processes to clean them before use, or discharge to the atmosphere.

Gas engine (obs) Very large gas engines, running on cleaned blast furnace gas, were formerly used in many plants for driving blast furnace blowers and for electricity generation. Very rarely, rolling mills were driven by gas engines.

Gas fuel Blast-furnace, coke-oven, and manufactured gas are all used as fuels in steelworks and natural gas is now used to a small extent as well.

Gas man　　The workman who attended to the gas producer used at reheating and open-hearth furnaces.

Gas port　　*See* Port.

Gas producer (Dowson producer) (Siemens producer)　　A cylindrical refractory-lined chamber for gasifying coal to produce gas as a furnace fuel. Gas producers are now virtually obsolete but were formerly used extensively with open-hearth furnaces. The gas is called producer gas to distinguish it from other gaseous fuels. A well-known gas producer was the Dowson from which producer gas was often known as Dowson gas. Another was the Siemens (the first); hence Siemens gas.

Gassed, gassing (of a mould)　　*See* Carbon dioxide process.

Gas strip (obs)　　Iron strip rolled specially for making into gas pipes.

Gas valve　　The valve controlling the admission of gas to a Cowper hot blast stove. It is used in conjunction with the chimney valve, hot blast valve, and cold blast valve to reverse the flow in the stove and so change from heating to delivering heat.

Gate (downgate) (git)　　The vertical passage leading to the mould cavity in a mould. There may be several on a single casting. Often spoken and written as 'git'.

Gathering (obs)　　A defect in pack-rolled steel sheets caused by a small area of the sheet sticking to the roll.

Gauge (G)　　A means of measuring the thickness of sheet or strip or the diameter of rod or wire. Many gauges existed, the best known being Birmingham Gauge (BG) and Imperial Standard Wire Gauge (ISWG or SWG). Thicknesses and diameters are now usually given in decimals and the old gauges are becoming obsolete, but the word itself is still used, especially in a general sense, eg light gauge sheet etc. Until recently sheets, hoops, and strip were usually measured in BG and wire in SWG.

Gayley process (obs)　　*See* Blast cooling.

Gero process (mould degassing)　　An American proprietary process for degassing molten steel in the ingot mould. A hood is fitted to the mould top and connected to a vacuum-pumping source. Metal is teemed through the hood via a small tundish which prevents air from being sucked into the mould.

Gidlaw puddler (obs)　　*See* Puddling machine.

Gig-tyre iron (obs)　　*See* Cart tyre iron.

Girder hammer   A steam hammer with the steam cylinder carried on a pair of girders fixed to two uprights. The girders had a wide span and so gave plenty of room for large forgings.

Git   *See* Gate.

Git cutter   A machine for cutting the gits or gates of castings during fettling.

Gjers kiln (obs)   A cylindrical kiln formerly used for calcining iron ore before smelting.

Gjers pit (obs)   The original form of soaking pit; a simple cavity lined with refractory in which ingots were placed while the heat soaked uniformly through them.

Glass hard   Steel at the highest degree of hardness it will take. It is then very brittle.

Glaze (obs)   Glazing the bottom was the working-up of a scrap ball in the puddling furnace which had been heavily repaired, or completely refettled. It consolidated the newly made bottom.

Glazed pig iron   *See* Blazed pig iron.

Gleed   A common name in the Midlands for small coke or breeze. Gleeds were used in smiths' hearths.

Glut (obs)   A wedge-shaped piece of iron welded into a corner of a wrought-iron forging where it had thinned down through being turned through a sharp angle, or to fill in any V-shaped cavity in a weld.

Glut iron (obs)   Wrought-iron bar rolled to a wedge-shaped cross section, in various sizes, for making gluts.

Gob up   A blast furnace is said to be gobbed up, or to gob up when the charge becomes solid in the crucible.

Goggle valve (spectacle valve)   A type of valve used as a stop-valve in gas mains. It consists of a piece of plate pivoted so that it can swing in a slot in the main. The plate has a large hole in one part, equal in diameter to the bore of the main. When this is swung in line with the main, it is open. When it is swung out the solid portion swings in, so closing the main.

Goose neck   The bent pipe connecting the bustle pipe or horseshoe main to the tuyere of a blast furnace. Also used for the pipe which takes the blast from the trunnion to the windbox on a Bessemer converter.

Gondola (rare)   A slag ladle, running on rails.

Gordon-Cowper Whitwell stove (obs)   A type of blast furnace

hot blast stove of the regenerative type but with its own
chimney on the top. More common in USA than Britain.

Gothic pass   A pass shaped like a double gothic arch in a pair
of rolling mill rolls.

Gothic section   A bar or billet of square cross section with
rounded corners and slightly convex sides.

Gouge   A tool similar in shape to a carpenter's gouge and used
by the blacksmith for cutting iron or steel hot to concave
shapes. It has a handle set at right angles, is held by the smith
and struck by the striker.

Grade of pig iron (obs)   *See* Number one (two etc).

Grade of wrought iron (obs)   *See* Wrought iron grades.

Graef rotor process   *See* Rotor.

Grain, come to   *See* Nature.

Grain flow   When steel or wrought iron is worked hot (by
rolling or forging) the crystal structure and the non-metallic
inclusions are elongated in the direction of working, so giving a
fibrous structure. The metal is always strongest along the grain
flow lines and weakest at right angles to them.

Grain roll   A rolling mill roll cast in sand and therefore having
a relatively soft surface. Used for roughing and in some finishing
mills. For finishing, especially sheets, chilled rolls, cast with a
chill on the barrel, which makes it hard, are used. The remainder
of the roll is not chilled. Chilled rolls can be clear or definite
chill (when the chill is clearly defined) or indefinite chill (where
the chill or hardness decreases gradually from the face inwards).

Grain tin (obs)   *See* Block tin.

Granulated slag   Blast furnace slag broken into small pieces by
pouring it, molten, into water. It is used as a concrete aggregate.

Granulation   A method of forming molten iron into granules.
It is poured into cold water in a controlled stream and solidifies
as particles up to about 1½ in maximum. Is sometimes used in
place of casting to pigs when the iron is to be remelted for
steelmaking because granules are easily handled mechanically.

Graphite   A naturally occurring form of carbon. It appears as
small flakes in grey cast iron.

Graphite crucible (obs)   A steelmaking crucible made from
natural graphite mixed with clay and sand instead of the more
usual clay mixture. Not used extensively.

Graphitic carbon   That part of the carbon content of cast iron which is present as graphite. Carbon is also present as combined carbon (dissolved in the metal). Graphitic and combined carbon added together equal the total carbon content (often written TC).

Grate bars   *See* Screen bars.

Gray iron   *See* Grey iron.

Grease pot (1) (obs)   A bath of molten palm oil or tallow or a mixture of both, into which sheets were dipped after hand tinning, to allow the surplus tin to drain off.

Grease pot (2) (obs)   That part of a tin pot on which the molten palm oil floated.

Greasy heat   *See* White heat.

Green   Broadly any material such as sand or refractory which has not been subjected to heat. Thus an unfired refractory brick is green. *See also* Green sand.

Greenawalt process   A process for converting small pieces of iron ore into sinter (qv). Sintering is done in pans, which are turned over mechanically to discharge the sinter.

Green sand   In a foundry green sand is a natural sand (often with a little clay) used in slightly damp form to make a mould. The name has nothing to do with the colour (most moulding sands are red) but refers to the 'green' or unbaked form in which it is used. In use the sand gradually turns black, partly from burning and partly from getting mixed with blacking (qv).

Green strength   The strength of moulding sand after moulding. It must be sufficient to withstand handling and if used green, to resist the wash of the molten metal when the mould is filled.

Grey cast iron   *See* Grey iron.

Grey iron (grey cast iron) (gray iron)   Cast iron in which all or most of the carbon is present as small graphite flakes. It is the common cast iron of commerce.

Grey mill   The original form of universal beam rolling mill, named after the designer. (*See* Universal mill.)

Grip die   Used in forging, a grip die holds the metal while it is upset. The die is split so that it can be opened to release the forging and closed down tightly to grip it beforehand.

Grog   Old, burnt fireclay, crushed fine and used to mix in with new clay for making refractories and, formerly, crucibles.

Groove   *See* Pass

Ground charger   *See* Charging machine.

Grouser section   A special rolled steel section produced for the tracks of crawler vehicles.

Guard   *See* Guide.

Gubbin (obs)   A type of ironstone formerly found in the Black Country.

Guide   Any bar, box, channel, or other contrivance to guide the piece being rolled into the appropriate pass in the rolling mill. Distinct from the guard (or stripping plate) which may be similar in form but is on the exit side and prevents the piece from collaring (qv).

Guide bar   *See* Cramp bar.

Guide mark (guide score) (guide scratch) (guide shearing)   A surface defect on rolled products caused by contact between the piece and the guide.

Guide mill   A rolling mill for small rounds, bars, or sections in which the final pass has a box guide, which completely surrounds the piece and both guides it and prevents it from twisting. Is particularly necessary when rolling from oval to round or diamond to square.

Guide score   *See* Guide mark.

Guide scratch   *See* Guide mark.

Guide shearing   *See* Guide mark.

Guillotine shear   A shear, usually used for cutting plate, sheet, and strip, in which the lower blade is stationary and the upper blade moves downwards to make the cut.

Guillotining   Cutting plate, sheet, or strip in a guillotine shear.

Gun   *See* Mud gun.

Gunite   A refractory mortar applied by a cement gun to repair a furnace lining.

Gunning   Applying refractory cement by a cement gun.

Gutter   A shallow depression round the impression in a drop-forging die to allow the excess metal or flash to flow out.

Hadfield manganese steel   The first manganese steel, made by Sir Robert Hadfield, containing typically 12 to 14% manganese, 0.95 to 1.4% carbon, and 0.3 to 1.0% silicon.

Haematite   *See* Hematite.

Half-and-half (obs)   *See* Hot-cold.

Half bloom (obs)   *See* Loop.

Half round   A rolled section which, as its name indicates, is half a true circle in cross section.

Hall's process (obs)   *See* Puddling.

Hallsworth system   A proprietary moulding system for high production rates. The moulding equipment is arranged round an automatic moulding machine.

Hammer cogging   *See* Cogging.

Hammer driver   *See* Shingler.

Hammer, drop   *See* Drop hammer.

Hammered bars   Bars of iron (obs) or steel which have been made to final shape and size under the hammer instead of being rolled. Some special steels are still made in this way when the quantity is small.

Hammer, helve   *See* Helve hammer.

Hammer lap   A folded-over piece of metal hammered into the surface of a forging. It is a defect, caused by bad forging practice.

Hammer scale   Scale driven off iron (formerly) or steel during hammering.

Hammer slag (obs)   Slag expelled from a puddled ball during

Hammer, steam

shingling to a bloom.

Hammer, steam   *See* Steam hammer.

Hammer, tilt   *See* Tilt hammer.

Hammer welding   *See* Forge welding.

Handbag (obs)   A pad of sacking, kept wet, and used to protect the hand of a worker in crucible steelmaking when handling hot tongs.

Hand-forged (smithy forged)   Components made by forging under the hand or power hammer, or using dies which do not completely enclose the forging.

Hand mill   Any rolling mill in which the piece is fed and manipulated by hand.

Hand rammer   A foundry tool used for ramming the sand into moulds. It usually has a rounded head at one end and a wedge-shaped head at the other.

Hand-rolled   A term applied especially to rounds rolled in a simple mill without box guides and with no oval-to-round passes, the piece being turned through 90° between passes by the roller, to prevent finning.

Hand rounds   *See* Hand-rolled.

Hand-shank   *See* Ladle.

Hanging   Sticking of the burden to the side of a blast furnace to form a scaffold. Sudden movement from hanging is a slip.

Hanging core (suspended core)   A core with a large print on top by means of which it can be suspended in a mould without any other support.

Hanker knife   *See* Opening.

Hard-drawn wire   Wire which has been subjected to a heavy drawing operation (reduced by more than 10% in one pass) and so has work-hardened.

Hardening   Increasing the hardness of steel by heat treatment. Mild steel (and wrought iron) will not harden; carbon and alloy steels will. The heat treatments are many and varied, depending on the chemical composition of the steel.

Hardening, selective   *See* Selective carburizing.

Hardie (hardy)   A short chisel-shaped tool with a peg to fit into the hardie hole of a smith's anvil. The metal to be cut is placed directly on the hardie and struck with the hammer to cut it.

Hardie hole (hardy hole)   *See* Anvil.

Hard tap   Any case of a blast furnace or steel furnace proving difficult to tap in the normal way is said to be a hard tap. Heavy drilling and burning out with oxygen may be necessary.

Hardy   *See* Hardie.

Harveyizing   A case hardening process for armour plate steel. The plate surface is covered with carbonaceous material, heated for about 120 h and then quenched by water jet.

Hazelett process   A method of continuously casting molten metal into slabs ready for rolling. The molten metal is poured between two moving horizontal (or almost horizontal) steel bands, cooled by large volumes of water on the sides opposite to the metal. The process is mainly used for non-ferrous metals but has been applied to steel.

Healing   The welding up of cracks in an ingot during the first few rolling passes.

Heart and square   A small smoother used by foundry moulders. It is heart-shaped at one end, square at the other, for reaching into pointed or square parts of a mould.

Hearth (1)   Strictly the bottom of a blast furnace lining but often applied to the crucible as well.

Hearth (2)   The bed or working portion of any furnace which holds the metal being heated or melted.

Hearth (3)   The coal or coke fire used by a smith to heat metal for forging.

Hearth casing   *See* Jacket (1).

Hearth, coke   *See* Coke hearth.

Hearth jacket   *See* Jacket (1).

Hearth staff   A blacksmith's term for a poker.

Heat   In iron- and steelmaking heat has a specialized meaning besides the normal one. It is (1) the steelmaking operation from charging raw materials to tapping the molten steel, (2) the working of wrought iron in the puddling furnace from charging pig iron until the last ball is taken out.

Heat colour   Steel changes colour as it is heated, and an approximate idea of the temperature can be gained by the colour.

| Incipient redness | = | approx | 525°C |
| Dull red | = | approx | 700°C |

Heat cracking

Cherry red	=	approx 900°C
Deep orange	=	approx 1,100°C
White heat	=	approx 1,300°C
Dazzling white	=	approx 1,500 to 1,600°C

Cf. temper colour.

Heat cracking   Steel or iron subjected to alternate heat and cold over a period will crack on the surface. Rolls used in hot mills suffer from this trouble.

Heater   The man in charge of a heating furnace, especially a re-heating furnace.

Heather   *See* Descale.

Heaton process (obs)   A means of making steel from cast iron, using nitrate in a type of converter rather like a cupola instead of atmospheric air. It was not widely adopted.

Heat-resisting irons   Alloy cast irons with a high nickel and chromium content, made specially to resist oxidation and retain their mechanical strength when used at high temperatures, eg as furnace components. They are akin to the stainless steels (qv).

Heat-resisting steels   Alloy steels with a high nickel and chromium content made specially to resist oxidation and retain their mechanical strength when used at high temperatures, eg in gas turbines.

Heat treatment   Any one of many methods of heating steel to give it certain desired properties, eg hardening, softening, tempering. Heating solely for hot working, eg rolling or forging, is not included in this definition.

Heat waste (fire waste)   Loss of metal caused by scaling when the metal is reheated for rolling, forging or other processing.

Heaver-over (obs)   In the Black Country particularly the catcher or back-man at a single-stand sheet mill.

Heaver-up (obs)   Black Country for the catcher or back-man at the forge train.

Heavy iron   A defect in the spangle (qv) of galvanized sheet, caused by excess iron in the zinc in the galvanizing bath.

Heavy metal   *See* Scruff.

Heel   A relatively small amount of molten metal left behind after tapping a furnace.

Hell (obs)   A long-obsolete term for a cupola (qv).

Helve (obs)   A heavy cast-iron hammer, pivoted at one end, lifted by cams and allowed to fall by gravity. The nose, or frontal helve, commonly used for shingling, had its fulcrum at one end, the cam ring at the other and the hammer face and anvil in between the two. The belly helve, used mainly for making forgings, had the fulcrum and hammer face at opposite ends and the cam ring in between. Sometimes called a trip hammer as was also the tilt hammer (qv).

Hematite (red hematite) (bloodstone) (brown hematite) (wood hematite) (kidney ore)   A high grade iron ore, Fe_2O_3. It exists in various forms such as red, kidney ore, brown, and wood hematite, and is sometimes called bloodstone. The older spelling is haematite.

Hematite iron (hematite pig)   Pig iron made from hematite ores. Not very common today, but formerly in wide demand for the acid Bessemer process.

Hematite pig   *See* Hematite iron.

Henderson process (obs)   A process for removing tar and ammonia from blast furnace gases where coal was the fuel.

HERF   *See* High-energy-rate forging.

Heroult arc furnace   A direct electric arc furnace with three electrodes, each of which is connected to one phase of a three-phase electric supply. Current strikes arcs between the electrodes and a charge of metal in the hearth, generating intense heat. This is basically the common arc furnace used for steelmaking and it is usually called simply an arc furnace; the name Heroult being dropped.

Hexagons   The common name for steel bars rolled or cold drawn with a hexagonal cross section.

HF furnace   *See* High-frequency furnace.

High carbon steel   *See* Carbon steel.

High duty iron   A generic term for alloy cast irons having special properties such as heat- or corrosion-resistance, wear-resistance, special electrical properties, etc.

High-energy-rate forging (HERF)   Forging, in shaped dies in a machine which gives a very sudden blow, by expanding gas or by controlled explosion in a cylinder.

High-frequency furnace (HF furnace)   An electric induction furnace (qv) in which the current is supplied at frequencies much higher than normal. Used for steel melting, especially small quantities of high-grade alloys.

High line   *See* Furnace bank.

High-speed steel (1)   For statistical purposes the industry defines as high speed steel any alloy steel containing tungsten with or without molybdenum, such that when the percentage content (by weight) of molybdenum is multiplied by two and added to the percentage content of tungsten the sum is not less than 12.

High-speed steel (2) (red-hard steels)   In broad terms an alloy steel capable of being intensely hardened, and used for engineers' cutting tools. Sometimes called red-hard steels.

High-strength steels   A generic term for low-alloy steels made specially for structural purposes.

High-top-pressure operation   A method of running a blast furnace with a higher than usual pressure in the top of the stack (normal = about 2 lbf/in^2 ; high = 10 to 15). It increases the density of the air in the furnace, enabling more air to be blown in and so increases the output; it reduces dust carry-over with the waste gases and improves the coke rate (qv).

Hire rolling   Rolling, for an agreed fee, material supplied by anyone other than the owner of the mill. Some mill owners use spare capacity in this way; others make a speciality of hire rolling. Hire rollers are quite common in the Sheffield area.

H-iron or H-beam (obs)   An iron or steel beam like a letter H in cross section. More correctly a joist. The British Standard joist is now obsolete, having been superseded by the British Standard universal beam (qv).

H-iron process   A process for reducing iron ore fines with hydrogen. Especially suitable for countries where coke is not available. Not used in Britain.

Hi-top   A substitute for tinplate, the surface coating being of chrome oxide instead of tin.

Höganäs process   A method of making sponge iron (qv) by direct reduction.

Hogger   *See* Booster ejector.

Hogging   Cutting off runners, risers, etc, from castings by flame cutting.

Hojalata y Lamina process (HyL process)   A Mexican process of direct reduction for reducing iron ore with reformed natural gas. Not used in Britain.

Holding furnace   Any furnace which is designed simply to keep metal, which has been heated elsewhere, at a particular temperature for short periods, eg when there is a delay in a rolling mill which cannot take a billet immediately.

Holding time   The period of time, in heat treatment or re-heating, during which steel is kept at a constant temperature to allow the heat to penetrate uniformly through the mass of metal. Also called soaking.

Hole (1)   The hole in a wire-drawing die.

Hole (2)   *See* Pass.

Hole (3) (obs)   *See* Pot furnace.

Hole (4)   The chamber in a soaking pit which actually receives the ingots. Some soaking pits have several holes.

Hole, charging (obs)   *See* Charging hole.

Holley bottom   *See* Plug (2).

Hollow   A thick-walled tube (made by various methods) which is the raw material for seamless tube manufacture.

Holloware   The complete assembly of runner (or trumpet), bottom plates etc used for conducting the metal to the moulds in uphill teeming.

Hollow fire (obs)   A coke-fired furnace formerly used for heating iron and steel for forging. Coke was burnt in a refractory-lined chamber and the products of combustion passed to the chamber in which the piece to be heated was placed. The piece was never in contact with the burning coke.

Hollow forging   Any forged hollow body, such as a boiler drum.

Homo iron (rare)   A name sometimes given to low carbon steel.

Hook   *See* Hooker.

Hooker (hook man)   A man or youth who assists the catchers at a hand mill when rolling heavy pieces. He uses a long hook-ended bar suspended by a chain to lift the piece and lead its end into or over the rolls as required.

Hook man   *See* Hooker.

Hoop   ⎱ Strip specially rolled, usually in gauge sizes,
Hoop iron (obs)   ⎰ for making barrel and other hoops.

Hopkins process   The original name for electro-slag refining (qv).

Hopper   *See* Bunker.

Horns   A fault in rolling sheet or strip when, through incorrect camber, uneven pressure or worn rolls, projections like horns appear at the ends of the rolled product.

Horse   *See* Bear.

Horseshoe firing   *See* Loop firing.

Horseshoe iron (obs) (shoe iron)   Self-explanatory but it is worth recording that special sections were rolled, both plain and fullered, ie with a groove which farriers would otherwise have had to put in by the smithing process called fullering.

Horseshoe main (obs) (pig's pudding)   The blast main on an open forepart blast furnace. It did not encircle the furnace completely but broke off at the front above the forepart. Cf. bustle pipe.
Also called in the Black Country the pig's pudding.

Hot-blast cupola   *See* Cupola.

Hot-blast furnace   *See* Blast furnace.

Hot-blast iron   Pig iron made in the blast furnace with hot blast, so called originally to distinguish it from cold-blast iron. All pig iron is now made with hot blast and the distinction is unnecessary.

Hot-blast man   *See* Stove minder.

Hot blast stove (1)   A refractory lined cylindrical structure used for providing the heated air for a blast furnace. It is of the regenerative type (qv).

Hot blast stove (obs) (2)   The first hot blast stoves were of the pipe pattern; air was passed through a pipe or pipes and fuel or gas was burnt outside the pipe(s) to heat the air. All are now obsolete; only the regenerative type is used. Neilson's was the first.

Hot blast valve   *See* Gas valve.

Hot box (1)   A heated mould used for forming foundry cores from heat-setting sand.

Hot box (2)   *See* Steckel mill.

Hot-cold (obs)   A blast furnace blown with a mixture of hot and cold blast; often one-third hot to two-thirds cold, but it varied. Also called blowing half-and-half. Iron so produced was often called coddled iron.

Hot extrusion   *See* Extrusion.

Hot galvanizing   Surface coating of steel (and formerly wrought iron) by dipping it in molten zinc.

Hot metal (blast furnace metal)   In iron and steelmaking the term hot metal is confined to molten pig iron or blast furnace iron.

Hot-metal ladle   The term is restricted to the large rail vehicle type ladles used for carrying molten iron from blast furnaces to pig casting machines or steelworks. The torpedo ladle (qv) is one type, but hot-metal ladles can be open topped vessels.

Hot-metal mixer   *See* Mixer.

Hot-metal process   Any steelmaking process in which the furnace charge consists wholly or partly of molten iron. When the whole charge is of cold metal the term cold-metal process is used.

Hot-metal shop   Any steelmaking shop in which at least part of the charge is molten iron. Any shop using cold metal only is called a cold-metal shop.

Hot mill   Any rolling mill used for hot rolling.

Hot-pack rolling (obs)   The hot rolling method formerly used for making sheets for tinplate. There were three main methods; the five-part, the four-part, and the three-part. The sequence for five-part rolling was: heat, bosh (dip in cold water to remove scale), roll singles, reheat, roll singles, double, reheat, roll doubles, double again, shear, reheat, roll fours, double again, shear, reheat, roll eights. The three-part sequence was: heat, bosh, roll in pairs, reheat, match (put two sheets together one on top of the other), roll, double, reheat, roll fours, double, shear, reheat, roll eights. The four-part sequence was: heat, bosh, roll in pairs, reheat, roll singles, double, reheat, roll doubles, double, shear, reheat, roll fours, match fours, roll eights.

Hot-rolled   Any steel which has been rolled hot.

Hot rolling   Any rolling in which the steel is heated before passing it into the rolls.

Hot saw   A circular saw used for cutting steel to length immediately after hot rolling, while it is still hot. If greater accuracy is needed a cold saw, which is much slower, is used.

Hot set   *See* Chisel.

Hot-short   *See* Red short.

Hot Steckel mill   *See* Steckel mill.

Hot tears   Cracks formed in ingots during cooling, by stresses in the steel.

Hot tinning   Surface coating steel (and formerly iron) sheets by dipping them in molten tin.

Hot topping (1) (anti-pipings)   A method of keeping the top of an ingot mould molten long enough to 'feed' or fill up shrinkage cavities formed as the ingot cools. Done by feeder heads and/or exothermic heads on top of the mould. A dozzle (qv) is one way of hot topping.

Hot topping (2)   Progressive reduction of the arc current in vacuum or electro-slag remelting towards the end of melting. It keeps the top of the ingot molten for feeding (see hot topping (1)).

Hot trimming   Removing the flash from a forging while it is still hot.

Hot working   Forming or shaping steel while it is hot. This includes forging, rolling, stamping, and extruding, though all these can also be done cold.

Hot-working steels   Steels made specially for tools which are to be used for working steel hot, eg drop-forging dies.

Housing (standard)   The frames, formerly cast iron but now cast steel, of a stand of rolls. Each stand comprises two housings with the rolls, chocks, screws, etc.

Housing screw   Mounted in the housings to bear on the top roll chocks and so make it possible to adjust the roll settings.

Housing window   The gap in the side of a housing through which access can be had to the rolls, or the rolls can be withdrawn for changing.

Hovel   Commonly used, especially in the Black Country, for a small hut which served as a shelter, mess room, and rest room for the men working on a blast furnace.

Hungry iron (obs)   Pig iron rich in silicon and phosphorus and needing extra oxides for their removal in the puddling furnace.

Huntsman process (obs)   The original crucible process (qv).

Huntsman steels (obs)   Steels made by the crucible process (qv).

Hydraulic mill   A rolling mill in which the roll gap is set and maintained by hydraulic cylinders instead of housing screws. The design lends itself particularly to automatic gauge control (qv).

Hydrochloric acid   A strong acid, formula HCl, used, diluted, for pickling (qv) steel.

Hydrogen   A gaseous element which, if present in some steels, can cause trouble, such as cracking.

HyL process   *See* Hojalata y Lamina process.

I

I Beam (obs)   An old term for a joist (qv).

IC   *See* Basis box.

Icicle   When a furnace roof is overheated and begins to melt, the refractory drips off in small stalactites or icicles.

IG   British Standard Wire Gauge (qv) is occasionally written this way.

Igniter (ignition hood)   A gas or oil-fired hood over the sinter strand at the commencement of travel, to ignite the sinter mixture.

Ignition hood   *See* Igniter.

Immersion pyrometer   An electrical instrument with a throw-away tip on the end of a long arm. The tip is immersed in molten steel and an electrical current is induced. This is amplified and indicated directly on a dial or scale as degrees of temperature. The tip is destroyed and is replaced after every immersion.
Cf. radiation pyrometer, optical pyrometer.

Imperial standard wire gauge   *See* British Standard Wire Gauge.

Inactive mixer   *See* Mixer.

In blast   *See* Blow in.

Incidental element   *See* Tramp element.

Inclusions   Non-metallic particles in steel. They are objectionable and in high-grade steels special efforts have to be made to exclude or remove them.

Indented bar (deformed bar)   Steel bar with indentations rolled in or with a ridged surface for use in concrete reinforcements.

113

Indirect arc furnace   *See* Arc furnace.

Indirect teeming   *See* Teem.

Induction furnace   An electric furnace, essentially a transformer, in which an electric current is used to generate a rapidly alternating magnetic field. The electric coils surround a container or crucible in which steel is melted by the induced current. Usually fairly small (up to about 10 tons) and used for high-grade alloys. *See also* High-frequency furnace.

Induction heating   A method of heating steel by passing it through an alternating magnetic field.

Induction stirring   A method of stirring molten steel in a ladle in a degassing chamber. A high-frequency electric current in a coil round the ladle stirs the metal and brings it all to the surface for degassing.

Inert gas melting   Melting steel in an electric furnace under an inert gas such as argon so that deleterious gases do not enter the melt. (Not common).

Ingoing side   *See* Entry side.

Ingot   Steel cast in a metal mould ready for rolling or forging. It is distinct from a casting which is not rolled or forged. *See also* Shapes of ingots.

Ingot buggy   *See* Chariot.

Ingot car   *See* Chariot.

Ingot chariot   *See* Chariot.

Ingot cleaner (obs)   The member of a crucible steelmaking team responsible for dressing off with hammer and chisel, the flash from ingots and generally cleaning the surface. He also took care of the ingot moulds.

Ingot iron   An iron of high purity produced by special processing to keep down carbon, manganese, and silicon. Armco (qv) is an example.

Ingot mould   A cast-iron mould made specially for casting steel ingots. They are of many sizes (eg a few hundredweights to 25 or more tons) and several shapes (eg square, slab, hexagonal) according to the use to which the ingot is to be put. The ratio of mould weight to ingot weight is usually about 1 to 1 (ie a mould for a 5 ton ingot will weigh about 5 tons).

Ingot shapes   *See* Shapes of ingots.

Ingot slicer   *See* Slicer.

Ingot stripper (stripper crane) (stripper) A mechanical device for extracting ingots from the mould when they are solid. Strippers are usually attached to an overhead crane, which carries the ingots away for further processing after stripping.

Ingot tongs Large, heavy tongs forming a part of an ingot stripper (qv).

Ingot tonnage A term used for expressing steel output (works, district, or nation). It is being replaced by crude steel (qv).

Injectant The practice of injecting fuel at the tuyeres of a blast furnace is growing, to save coke. The fuels used are fuel oil (principally), methane, natural gas (not at present in Britain), powdered coal, and coke oven gas. They are known collectively as injectants.

Inoculation Treatment of molten cast iron in the ladle, or as it runs into the ladle, by the addition of specific quantities of special powdered materials to modify its composition or metallurgical condition.

Insulating bricks Refractory bricks used to reduce heat radiation from a furnace.

Intensifier *See* Steam-hydraulic press.

Interlocking piles *See* Sheet piles.

Intermediate annealing *See* Interstage annealing.

Intermediate train A group of rolling mill stands between the roughing and finishing trains of a continuous mill.

Interstage annealing (intermediate annealing) Annealing steel between production process (eg between cold rollings or drawings) to remove the effects of work hardening.

Investment casting (lost wax process) (precision casting) (cire perdue process) A modern (wartime) revival of an ancient process formerly used for casting statuary etc. A wax pattern is made and then 'invested' or coated with refractory slurry. On baking the coating, the wax melts out and a cavity is left the exact size and shape of the pattern. This is then filled with molten metal.

In-walls (obs) The lining of a blast furnace, particularly one of masonry or brick construction.

Iron (1) (n) A metallic element, chemical symbol Fe, specific gravity 7.87 at $20°C$, atomic weight 55.85, melting point $1535°C$. It has an affinity for oxygen and will alloy with many other elements, some of which can be beneficial. In general terms there are three commercial forms of iron: cast iron, an alloy of iron, carbon, phosphorus, silicon, and manganese, carbon being the primary alloying element; wrought iron, the commercially pure form of iron; and steel, which is an iron alloy with many other elements and forms a complete 'family' of metals.

Iron (2) (v)   If a blast furnace tuyere gets a deposit of iron and/ or slag on its nose it is said to have ironed, the building-up being known as ironing.

Iron (3)   *See* Gagger.

Iron (4)   *See* Sheet bar.

Iron granulation   *See* Granulation.

Iron notch (rare)   *See* Taphole.

Iron ore (stone) (mine)   Any one of the many oxides or carbonates of iron occurring naturally and used for smelting to iron. The term ironstone is often loosely applied to various kinds of ore but should be confined to the argillaceous or clay-bearing ores.

Ironstone   A general name for iron ore (qv) though it should really be restricted to the lower commercial grades.

IRSID continuous steelmaking process   *See* Continuous steelmaking.

ISWG   *See* British Standard Wire Gauge.

IX   *See* Basis box.

J

Jacket (1) (hearth jacket) (hearth casing) The sheathing, formerly of iron and now of steel plates, of the lower part of a blast furnace, below the lintel plate.

Jacket (2) (mould jacket) A metal or wooden frame placed round a mould to keep it together after removal of a snap flask (qv).

Jack star *See* Rumbling star.

Jam bar (obs) A curved-ended iron bar used at a puddling furnace during fettling, melting down, and boiling.

James spun cast process A variant of the centrifugal casting process (qv), in which the moulds are kept at a temperature of 300 to 350°C, to prevent chilling of the castings. It is used for automotive cylinder liners and piston-ring 'pots', which can then be machined in the as-cast condition.

Jar ramming *See* Jolt ramming.

Jet tapping *See* Explosive tapping.

Jig A metal frame used to check the dimensional accuracy of a casting. It is provided with contact faces at all the important points; if the casting is correct it will fit in the jig.

Jigging bar bank (Black Country) A bar (or light section) cooling bank in which the pieces are moved laterally from the mill exit to the cold side in a series of short steps by machinery.

Jobbing foundry Any foundry which makes small numbers of non-repetition castings to order.

Jobbing mill *See* Merchant mill.

Joist Iron or steel rolled so that its cross section resembles a

Jolt ramming (jar ramming)

letter H. Often called H-iron or H-beam. Iron joists are obsolete. Steel joists are often referred to by the abbreviation RSJ.

Jolt ramming (jar ramming)   A moulding technique in which the moulding box, pattern, and sand are carried on a machine table which is raised and dropped a number of times; the weight of the sand acts as its own rammer. Jolt moulding machines are often combined with a mechanical squeezing plate which comes down and completes the ramming; the machine is then called a jolt-squeeze machine.

Jolt-squeeze machine   *See* Jolt ramming.

Jones puddler (obs)   *See* Puddling machine.

Jumper steel   Special steel (usually 0.60 to 0.65% C, 0.20 to 0.40 Mn) for making the jumpers or bits of compressed-air rock drills.

Jumping   *See* Upset.

Jumping the blast   Varying the pressure of the blast on a blast furnace rapidly or taking it off and putting it back suddenly in an attempt to clear a scaffold.

Jumping three-high mill   A Swedish three-high rolling mill in which the pass-line remains constant and the complete stand moves up and down between passes.

Jump mill (obs) (Welsh mill)   A two-high sheet mill (especially Welsh sheet and tinplate) in which the top roll was free to rise in the housings and 'jumped' up every time a piece passed through. The top roll was not driven.

Jump weld   Welding two ends of a bar without scarfing, eg as in light chainmaking. Cf. scarf.

Junghans process   A pioneer continuous casting process (qv).

K

Kahlbaum iron   Iron produced in Germany to a purity of 99.975%. It is of laboratory interest only.

Kaldo process   A Swedish steelmaking process. Molten iron is charged into a refractory lined vessel which can be rotated at up to about 30 rev/min and oxygen is blown in through a water-cooled lance. The molten iron is rolled round by the rotating vessel and the surface is continuously changed, so that it comes into contact with the oxygen and is very rapidly decarburized and dephosphorized. A 30 ton charge of iron can be refined to steel in about 40 min. The vessel can be turned vertically for charging and below horizontal for discharging or tapping.

Keeper (furnace keeper)   The man in charge of tapping and stopping a blast furnace and of the general management of the forepart. Sometimes used loosely; the duties varied from place to place.

Keller furnace (obs)   An electric furnace of French origin adapted for iron smelting in two stages.

Kelly converter   A converter virtually the same as the Bessemer (qv) invented in America at about the same time.

Key   A wedge for fastening hammer faces or tools into the tup (qv) and anvil.

Key section   *See* Specials.

Kick   An explosion between the two bells of a blast furnace.

Kicker   *See* Kick stamp.

Kick-off   A mechanical device for pushing plates, sheets or other rolled products, sideways off a run-out line for disposal at points other than the end.

Kick stamp   A light drop hammer in which the tup was connected to a rope running over a pulley and down to near ground level, where it terminated in a stirrup. The hammer was operated by foot and releasing it to let the tup fall. Hence the name of the operator — a kicker — which was transferred to the driver of a power drop stamp. Sometimes the pulley was mechanically driven.

Kidney ore   A form of hematite iron ore which is so called because it is kidney shaped.

Killed steel   *See* Killing.

Killing (deoxidation)   Removing oxygen from molten steel by adding, immediately before casting, elements such as aluminium, manganese, or silicon, which have an affinity for oxygen, with which they unite and pass into the slag. Manganese and silicon are usually added as alloys with iron, eg ferro manganese, ferro silicon, or spiegeleisen. Steel so treated is called killed steel. Insufficiently killed steel is called underkilled, and that not killed at all is unkilled.

Killing fire (obs)   Some crucible steelmakers used to give an extra fire (qv) after the steel had been melted, to superheat the metal ready for teeming.

Kiln   A shaft furnace used for roasting or calcining ores at relatively low temperatures. Similar to some kilns used in other industries, eg lime kilns.

Kink   *See* Buckle and kink.

Kish   Solid graphite which has escaped from molten iron during melting, and floats about in the air in the form of tiny flakes.

Kiss core   A core without prints.

Kisser   A fault in sheet steel. It is a patch of scale remaining on the sheet after pickling and is caused by two sheets being partly in contact so that the acid does not reach that spot.

Kjellin furnace (obs)   An electric furnace of Swedish origin. It was of the induction type.

Kling ladle   A type of ladle, spherical in shape, used for receiving molten iron from the blast furnace, and transporting it to the steelworks.

Knobble (obs)   *See* Nobble.

Knocker-up (obs)   A hand tool somewhat similar to a rabble, used in fettling a puddling furnace.

Knockout (shake out)   The area of a foundry where castings are knocked out of the moulds when solid (but not necessarily cold). In a mechanized foundry the knockout itself is mechanized, the moulds being passed over a shaking grid, through which the sand falls.

Krupp-Renn process   A German method of direct reduction of iron ore with coke breeze in a revolving kiln.

Krupp's disease (obs)   *See* Temper brittleness.

L

Lacquered plate   Tinplate coated on one or both sides with lacquer for extra protection or for appearance.

Ladle   A bucket-shaped vessel, refractory lined, for carrying molten metal. Can be of any size from a hand ladle (in a foundry often called a hand shank or shank) carrying a few pounds to a crane ladle carrying several hundred tons.
See also Teapot ladle, Lip-pour ladle, Bottom-pour ladle, Torpedo ladle.

Ladle additions   Alloying elements added to a steel melt after it has been tapped into a ladle.

Ladle analysis   The average chemical analysis of a heat of steel just before it is tapped into the ladle.

Ladle degassing   Degassing of molten steel while it is held in a ladle. The ladle may be put in an evacuated chamber.

Ladle-to-ladle degassing (Stream degassing)   Degassing liquid steel by teeming it from a ladle into a second ladle, the latter being in a vacuum chamber. A second, or pony ladle is sometimes mounted on the vacuum chamber and connected to it through a vacuum seal. The first ladle then discharges via the pony ladle.

Lamination   A fault in rolled steel, caused by non-metallic inclusions or other discontinuities which cause the metal to split into layers.

Lancashire hearth (obs)   *See* Finery.

Lancashire iron, Swedish (obs)   *See* Swedish Lancashire iron.

Lance   A pipe used for introducing a gas into molten metal. It

Langloan process (obs) (Addie process)
is usually water cooled. The oxygen lance used in oxygen steelmaking is an example.

Langloan process (obs) (Addie process) A process for recovering tar and ammonia from the gas from a coal-fired blast furnace.

Lap (shut) (overlap) A fault in rolled products, caused by a piece of the metal being folded over on the main body and failing to be welded up again.

Lapper (obs) (cold roll lapper) A fault in cold-rolled sheet for tinplate, caused by part of a sheet overlapping another.

Lap weld A weld in which a piece of one of the parts being joined overlaps the other.

Larry car A rail-mounted carriage with bottom discharging hoppers running across the top of a coke oven battery for charging the ovens with coal.

Lattins (obs) *See* Sheet (2) and Sheet sizes.

Launder *See* Spout.

Lauth mill A three-high rolling mill, the middle roll being smaller than the other two, which are driven, while the small roll is idle. Used for plate and sheet rolling but now rare.

Laying up (obs) Building up wrought iron slabs by welding, on the end of a porter bar (qv), in preparation for making a large forging, eg a ship's propellor shaft. The slabs were first heated in pairs in a furnace, separated by small pieces of iron, called todgers, to allow the heat to get to both faces to be welded.

LDAC process Similar to the LD process (qv) but powdered lime is blown in with the oxygen to refine high-phosphorus irons.

LD process An oxygen steelmaking process of Austrian origin, now in world-wide use. Molten iron is poured into a vessel superficially resembling a Bessemer converter, but with a solid bottom, and refined by blowing oxygen through a water-cooled lance entered through the mouth of the vessel.
The abbreviation LD is derived from Linzer-Düsenverfahren (Linz jet process) which was the original Austrian name, or from Linz and Donawitz, the two towns where it was first used. LD is now universal.

Leaded steels Carbon, alloy, or stainless steels containing a small amount (0.15 to 0.35%) lead, which is added to make the steels more easily machinable.

Leader pass   The pass before the finishing pass in a rolling mill, eg the oval before the round or the diamond before the square. The pass before the leader is the strand pass.

Lean gases   Gases, such as blast furnace gas, which are low in calorific (heating) value.

Lean ores   Iron ores which are of low quality, ie low iron content.

Lectromelt furnace   A proprietary electric furnace.

Ledloy steel   A proprietary free-cutting steel of American origin, also made in Britain.

Length-to-width rolling   *See* Broadsiding.

Leopard spots   Faults in the form of grey spots on tinplate.

Let down   *See* Temper (1).

Leveller (roller leveller)   A multi-roll machine which flattens steel plate or sheet being passed through it by reversed bending.

Levelling   Flattening sheet or plate.

Lid   *See* Pot lid.

Lie back   If some of the iron does not come out readily when a blast furnace is tapped, it is said to lie back.

Lies   *See* Lyes.

Lifter   *See* Gagger.

Lifting table   *See* Tilting table.

Lime dip (liming)   A treatment given to steel bars and wire to prepare them for drawing. The lime neutralizes traces of pickling acid and seals the oxide coating.

Limestone   Calcium carbonate ($CaCO_3$). Used as a flux in iron and steelmaking, it combines with impurities to form a slag.

Limewash   Limewash is sprayed on the moulds of a pig-casting machine to prevent the molten metal from sticking to the cast-iron moulds.

Liming   *See* Lime dip.

Limiting angle of bite   *See* Bite.

Limonite   A form of iron ore, $2Fe_2O_3 3H_2O$. Contains up to 60% Fe.

Lines   The internal shape of a blast furnace, seen as a vertical cross section when it is first lined or relined. After a campaign the lining is burnt and damaged; in this condition the internal shape is called the blowing-out lines.

Lining   The refractory layer on the inside of any furnace or ladle. Its nature varies with that of the metal being heated or melted.

Liningless cupola   A foundry cupola without an orthodox refractory lining. The steel shell is cooled by large volumes of cold water cascaded down it.

Lintel (mantel) (mantle)   The steel ring which stands on the columns of a blast furnace and supports the stack.

Linz-Donawitz process   *See* LD process.

Linzer-Düsenverfahren   *See* LD process.

Lip-pour ladle   A ladle used in foundries. The metal is poured over a lip at the top by tilting the ladle.

Liquid fuels   Various grades of oil fuels, from crude oil down, are used as steelworks furnace fuels nowadays to the virtual exclusion of coal, but natural gas is now being used in place of oil.

L-iron (rare)   Another name for angle (qv).

List (obs) (wire)   The small bead of tin left on the lower edge of a hand-dipped tinplate as it was removed from the tinning pot. It was itself removed in a subsequent operation in a small pot of molten tin, the list pot.

List pot (obs)   *See* List.

Live pass   *See* Pass.

Live roller   *See* Roller table.

Loam board   *See* Loam moulding.

Loam moulding (sweep moulding)   Usually pronounced 'loom'. Moulding in a mixture of sand, clay, straw, and horse manure or other binder, used as a wet, stiff slurry pasted over a former and strickled or struck up by a shaped strickle or loam board to the shape required. Moulds and cores can be made in this way. Now uncommon.

Long weight   *See* Short weight.

'Loom'   *See* Loam moulding.

Loop (obs) (loup)   The piece of wrought iron from the finery before it was hammered to form a half bloom which itself after reheating and hammering became a bloom. This had a small square knob (the ancony) at one end and a larger one (the mocket head) at the other.

Loop firing (horseshoe firing)   A method of firing a soaking pit (qv). Burning fuel enters through a port in the upper part of one end wall, describes a loop and exhausts through a port at the lower part of the same end wall. Cf. umbrella firing, straight-line firing.

Looping mill (Belgian mill)   A hand rolling mill in which, as a piece emerges from one stand it is seized by an operator with tongs and fed into the next pass. It is thus in two or more passes at the same time and forms a loop between the passes. Looping is often mechanized by means of a kind of trough called a repeater.

Loose pattern   Any casting pattern which is not designed for use in a moulding machine, ie a pattern for hand moulding.

Loose piece (false piece)   A part of a pattern which can be detached to enable the pattern to be withdrawn.

Lost wax process   *See* Investment casting.

Loup   *See* Loop.

Low-carbon steel   *See* Carbon steel.

Low Moor iron (obs)   A very high quality wrought iron formerly made at Low Moor Ironworks, Bradford.

Low-shaft furnace   A blast furnace with a short (or low) stack, designed to use low grade ore and bituminous coal. It is little more than experimental at present.

Lug   A projection on the side of a moulding box. There are several on each cope and drag and they are held together while casting with clamps.

Luhrmann front (obs) (Lurmann)   *See* Forepart.

Lute (v)   To seal the joints between an annealing box and its cover (or a crucible and cover) by means of a plastic mixture of fireclay and other materials.

Lyes (obs) (lies)   Water in which bran had been steeped for some time. It was used for the first stage of pickling in tinplate making.

Lying pipe (obs)   The horizontal pipe or main in a pipe-type hot blast stove.

Machine (obs) In a Black Country ironworks always a weighing machine. Probably little used elsewhere.

Machine man (obs) The man in charge of a weighing machine.

Machine cast pig (motherless pig) (sandless pig) (chill-cast pig) Pig iron cast in iron moulds, moved by machine past the pouring point. The line of moulds is called the strand. All pig is cast in this way today. A rather rare term is motherless pig.

Machine moulding Production of foundry moulds by any combination of mechanical equipment.

Machine puddled iron (obs) Wrought iron made in a mechanical puddling furnace, of which there were several, none very successful.
See Puddling machine.

Macro *See* Macrograph.

Macrograph A reproduction, photographic or otherwise, of an object (in this context an iron or steel surface) not magnified more than × 10; often actual size. A sulphur print (qv) is an example. Often abbreviated to 'macro'. Cf. micrograph.

Madaras process A batch process for making sponge iron from iron ore, using reformed natural gas as the reducing agent.

Made-up nozzle The nozzle of a bottom teeming steel ladle is said to be made up if the steel is too cold and solidifies in the nozzle. It can be opened by burning out (ie using an oxygen lance).

Maerz furnace A modified open-hearth furnace (qv) in which the front and back wall and roof are made of prefabricated units, easily replaceable for repairs.

Magnesite   Essentially magnesium carbonate ($MgCO_3$) but now generally applied to materials containing magnesia (MgO). It is used as a basis for high-grade refractories.

Magnetic comparator (magnetic sorter)   A means of sorting steel bars. There are two magnetic coils, one of which is placed over a standard bar. Other bars are passed under the second coil; if there is any difference in magnetic permeability this is indicated by instruments and the wrong or doubtful bar can be rejected.

Magnetic crack detection   A non-destructive test for revealing cracks in steel. A light oil containing iron oxide particles is washed over the surface and the steel is subjected to a magnetic field. If there is a crack north and south poles will form and these will show up as a pattern in the iron oxide. There are several variations, including some proprietary ones.

Magnetic oxide   On of the forms of iron oxide, Fe_3O_4. Cf. ferric oxide and ferrous oxide.

Magnetic separation   Some iron ores are freed of many impurities by passing them over a large electromagnet designed to attract the ore, which is magnetic, and send it in one direction, while the non-magnetic impurities go the opposite way.

Magnetic sorter   *See* Magnetic comparator.

Magnetite   Magnetic iron ore, FeO,Fe_2O_3. Contains up to 72% Fe.

Magnet steels   Highly specialized alloy steels (some very highly alloyed) made, as the name indicates, for the production of permanent magnets. Their manufacture is limited to a few specialist firms.

Make   *See* Yield.

Make a bottom (1) (obs)   Of a puddling furnace. *See* Bottom.

Make a bottom (2)   To prepare and fettle an open-hearth furnace hearth.

Malleable cast iron   A form of cast iron converted by heat treatment into a strong ductile material. *See* Whiteheart iron, Blackheart iron, Pearlitic malleable iron.

Malleable iron (1)   *See* Malleable cast iron.

Malleable iron (2)   An old name sometimes given, confusingly,

to wrought iron (qv). It should be confined to malleable cast
iron — the two materials are quite different.

Malleablizing Converting cast iron to malleable cast iron.

Malleable pig iron Blast furnace iron of a composition suitable
for making malleable cast iron.

Man-cooler A large electric fan, usually portable, used to direct
a current of air on men working in very hot places.

Mandrel A solid steel bar used to support a hollow forging
while it is under the hammer or press.

Mandrel forging Producing a hollow forging on a mandrel.

Manganese An element used in steelmaking. *See also* Ferro-
manganese.

Manganese steel An alloy steel containing various percentages
of manganese.
High-manganese steels (up to about 14% Mn) are very
resistant to wear.

Mangle (1) (v) To flatten plates in a mangle. *See* Mangling.

Mangle (2) (n) A roller flattening machine for plates. *See*
Mangling.

Mangling Flattening steel plates by passing them cold
between multiple rollers. It is similar to roller levelling but the
term is usually confined to heavier plates.

Manipulator (1) (forging manipulator) A power-operated car
with tongs capable of holding, turning and lifting an ingot or
forging under the forging press. The manipulator may run on
rails or on road wheels and may be electrically powered from a
trailing cable, or have its own diesel engine.

Manipulator (2) Heavy side plates set at right angles to and
near the rolls of a heavy rolling mill. They can be moved
sideways mechanically to guide the piece being rolled into the
correct position for rolling.

Manometer *See* U-tube.

Mantel *See* Lintel.

Mantle *See* Lintel.

Maraging steels A range of nickel alloy steels of recent
development capable of simple heat treatment to produce very
high strength and toughness.

Mark *See* Brand.

Marked bars (obs) *See* Brand.

Market wire   Bundles of low carbon steel wire made from basic open-hearth or basic Bessemer steel and sold in specified weights, usually about 100 lb but sometimes 63 lb.

Marking   *See* Brand.

Martensitic stainless   *See* Stainless steel.

Martien process (obs)   A method of decarburizing molten cast iron by running it through a trough or launder in which there were several small holes through which air or steam was blown. It was similar in principle (though not in practice) to the Bessemer process, with which it was contemporary, but it was not successful.

Martin-Siemens   A rare reversal of the term Siemens-Martin process (qv).

MA steel (obs)   A term sometimes applied to basic steel produced in a side-blown converter (qv).

Mast   The vertical standard which carries, by means of a supporting arm at right angles, the electrodes of an arc furnace.

Matching (1) (obs) (scraping)   Scraping the scale off rolled hoops with a scraper operated by a boy, prior to the final pass.

Matching (2) (obs)   Placing iron or steel sheets together to form a pack, in the pack-rolling process.

McKee distributor   *See* Distributor.

Medium carbon steel   *See* Carbon steel.

Meehanite   A proprietary cast iron with improved physical properties. There are several grades.

Melingriffith machine (obs) (Melingriffith pot)   A combined pickling and tinning machine named after Melingriffith tinplate works, Wales, where it was invented.

Melingriffith pot   *See* Melingriffith machine.

Melt (1) (v)   To reduce metal to liquid form either for casting or for refining or converting to another form. Not to be confused with smelt (qv).

Melt (2) (n)   The molten charge of metal in the furnace.

Melter   A generic term for anyone normally engaged in melting metal, but particularly applicable to steel melters. It should be noted that a steel melter was in the past sometimes called, incorrectly, a 'smelter' but the usage is now rare and should be discouraged, since melting and smelting are quite different operations.

Melter, first-hand   *See* First-hand melter.

Melter, second-hand   *See* First-hand melter.

Melters (obs)   An old name for foundry pig iron.

Melter, steel   *See* Melter.

Melter, third-hand   *See* First-hand melter.

Melting down   That part of a steelmaking process after charging and before refining (qv), when the action is purely one of melting.

Melting point   *See* Freezing point.

Menders   Tinplates which are imperfect but which can be recoated or otherwise rectified to full or second quality. Cf. primes, wasters.

Merchant iron (obs)   Common but not precise. It really refers to the common rolled wrought iron sections such as rounds, squares, hexagons, flats, which would be stocked by an iron merchant. But what a merchant stocked depended on his locality and the trades he served, so merchant iron could refer to many different sections.

Merchant mill (merchant train) (jobbing mill)   Any rolling mill doing a general jobbing trade.

Merchant train   *See* Merchant mill.

Mesh-belt furnace   A heat treatment furnace with a woven wire continuous belt running right through it. Articles are placed on the belt, either loosely or in containers, and the belt, mechanically driven, takes them through the heating zones.

Metal (obs)   *See* Refined iron.

Metal carrier (obs)   *See* Pig lifter.

Metal mixer   *See* Mixer.

Meteoric iron   The only form of native iron, of meteoric origin. Its composition varies but there is usually a fairly high percentage (up to 20%) of nickel. It is only a curiosity.

'Micro'   *See* Micrograph.

Micrograph   A graphic reproduction of an iron or steel surface magnified more than ×10; usually much more. If done photographically (as is usual) it is called a photomicrograph. Often abbreviated to 'micro'. Cf. macrograph.

Micrometer (flying)   *See* Flying micrometer.

Micron   *See* Torr.

Middle-bottom   *See* Top-middle.

Middler   The front man at the middle stand of rolls in a finishing hand rolling mill.

Middling   Finishing the middle of a forging before the ends. This is done for a variety of reasons, not the least of which is that if reheating is required, it is easier to stick an end into a furnace than the middle.

Mike (flying)   *See* Flying micrometer.

Mild steel   Steel containing between 0.10 and 0.20% carbon.

Mill   Strictly any kind of rolling mill but often applied in the plural to a group of mills, the building housing them and the ancillary plant such as furnaces, In the USA the term steel mill can mean a complete steelworks.

Mill edge   *See* Edge (2).

Mill furnace   A reheating furnace, usually reverberatory, used for reheating piles and billets for rolling. Also called a ball furnace, but incorrectly as this term should be restricted to a furnace used for reheating wrought iron scrap balls.

Millitorr   *See* Torr.

Mill modulus   A measure of the stiffness or resistance to rolling load (which tends to separate the rolls) in a rolling mill. It is usually expressed in terms of tons load per inch of deflexion.

Mill output (theoretical)   *See* Theoretical mill output.

Mill pile   *See* Pile (1).

Mill scale   An oxide of iron formed on the surface of steel during hot working.
Cf. roll scale.

Mill size   *See* Size of rolling mills.

Mill streaks (obs)   A fault in hot hand-rolled sheet caused by ashes or dust being rolled into the surface.

Mill train   *See* Train.

Millwright   A workman responsible for installing, maintaining, and repairing iron and steelworks machinery.

Mine   Iron ore. If calcined called burnt mine.

Mine filler (obs)   *See* Box filler.

Mineral wool   *See* Slag wool.

Minette ore   An iron ore occurring in Lorraine, Luxembourg, Belgium, and northern France. It is of the limonite type.

Mismatch (offset)   A fault in drop forging, caused by the impressions in the top and bottom dies getting out of line.

Misrun   A casting defect caused by the molten metal failing to fill the mould fully and correctly.

Mitre (obs)   A faggot of round iron bars arranged round a central bar, ready for faggoting.

Mixed blast   A modification of the Bessemer process, using some other gas for blowing as well as air. The VLN process (qv) is the best example.

Mixed fuels   Mixtures of coke-oven and blast-furnace gas are sometimes used as furnace fuels in steelworks.

Mixer (metal mixer) (hot metal mixer)   A large refractory lined vessel used to store temporarily iron from the blast furnace and on its way to steel furnaces. It enables the metal from several casts to be mixed, and acts as a reservoir so that iron is always ready when wanted. There are two basic types; the inactive mixer, in which no adjustment is made to the composition of the iron; and the active mixer, in which some part of the steelmaking can be carried out.

MKW mill   A German proprietary four-high cold strip mill with a unique roll configuration.

Mocket head (obs)   *See* Loop.

Model (obs)   An old name for a pattern (qv).

Molochite   A refractory derived from china clay and used as a blast furnace stack refractory. Its principal constituents are silica (SiO_2) and alumina (Al_2O_3).

Molybdenum   A metallic element, symbol Mo, used in high-strength alloy steels.

Moly steels   (Pronounced molly). A general term for alloy steels containing molybdenum.

Monkey (1)   *See* Bleeder.

Monkey (2) (obs)   A small pile of firebricks arranged by the puddler on the firebridge of a puddling furnace to deflect the flames towards the working door.

Monkey (3) (obs)   The flame at the top of a puddling furnace stack when the furnace was on full heat.

Monkey (4)   The inner one of three concentric coolers of a slag notch, and the one which receives the plug.

Monkey (5)   *See* Slag notch.

Monkey tuyere   *See* Tuyere

Monolithic lining   A furnace refractory lining which, instead of being built of bricks, is made in one piece by casting, ramming, or gunning in situ.

Mood   *See* Mould (2).

Morgan mill   An American continuous strip or bar mill, named after the designer. Made and used in many parts of the world.

Morgoil bearing   A proprietary bearing for roll necks. Oil is fed under pressure to the bearing and there is no metal-to-metal contact.

Moss   *See* Branning.

Motherless pig   *See* Machine cast pig.

Mottled iron   Pig iron of medium silicon content, in which about half the total carbon occurs as graphite while the rest is combined. It is very hard and brittle. The name comes from the appearance of the fracture. *See also* Number one (two etc).

Mould (1)   A cavity formed in sand, loam, or other material, of the shape desired for a casting. Molten metal is poured in and allowed to solidify, after which the mould is broken open and the casting removed.

Mould (2) (mood)   A rough forged shape used as the basis for forging edge tools. Mood is a Sheffield variant.

Mould board   A board on which a foundry pattern is placed while the mould is made.

Mould cavity   The impression left in a foundry mould after the pattern has been removed.

Mould clamps   Metal clamps used to hold the mould cope and drag (*see* moulding box) together while casting.

Mould coating (mould dressing) (mould wash)   Any substance applied to the surface of a foundry mould to improve the surface finish of the casting. It can range from lampblack, derived from a smoky oil lamp flame, to proprietary liquids.

Mould degassing   *See* Gero process.

Mould dressing   *See* Mould coating.

Moulder   The craftsman who prepares moulds in a foundry.

Moulder's rule (shrink rule) (contraction rule)   Castings contract on cooling so the mould, and therefore the pattern, has to be made oversize to allow for the contraction or shrinkage. A special rule is used, with the graduations expanded to incorporate the allowance. Various metals contract differently, so different shrinkage allowances are used and the rules are graduated accordingly.

Moulding box (flask)   A box of metal or wood, open top and bottom, in which a sand mould is made for casting. Can be square, rectangular, or a special shape and is usually made in two or more matching parts. A complete box consists basically of two parts, the lower or drag and the upper or cope, but there can be three or more parts.

Moulding hole   *See* Casting pit.

Mould, ingot   *See* Ingot mould.

Mould jacket   *See* Jacket (2).

Mould wash   *See* Mould coating.

Mouth   *See* Throat.

M stockline   *See* V stockline.

Muck (obs)   A Sheffield term for the debris produced when breaking up a crucible furnace for rebuilding.

Muck bar (obs) (puddled bar, obs) (puddler's bar, obs)   The product of the first rolling of a puddled bloom at the forge train. It was very rough, with ragged edges, and was never sold, but always cut up, piled and reheated and rerolled at least once.

Mucky hole (rare)   A blast furnace taphole from which the iron does not run freely because it is too pasty.

Mud gun (taphole gun)   A pneumatic, hydraulic, or (formerly) sometimes steam, cylinder used for shooting a refractory plug into the taphole of a blast furnace to stop it after tapping. Also called clay gun.

Muffle furnace   A furnace in which the heating chamber is completely sealed off from the fuel. A cementation furnace was really of this type.

Muller   A machine for mulling foundry sand.

Mulling   A method of mixing foundry sand by means of a heavy roller. The action is a combination of rubbing and stirring.

Multiple moulds (stack of moulds)   A composite mould made of a stack of separate moulds arranged vertically, with a common downgate.

Multi-strand rolling   Rolling of two or more pieces side by side, in separate passes, simultaneously in a finishing mill, especially a continuous mill.

Mushet's special steel   *See* Mushet's steel.

Mushet's steel (Robert Mushet's steel) (Robert Mushet's special

steel) (RMS)   The original air-hardening steel for engineers'
cutting tools, invented by Robert Mushet. It is an alloy
containing about 2% carbon, 2% manganese, and 7% tungsten.

Mush zone   *See* Prestressed mill.

Music wire   Carbon/manganese/silicon wire used principally for
making small springs. Sometimes called piano wire (qv) though
the latter is really of higher grade.

'Nail' (obs)   *See* Anneal (2).

Nail rod (obs)   Small rolled rounds or slit squares from about $\frac{3}{16}$ to $\frac{3}{8}$ in formerly made in iron or steel for hand wrought nail makers. Nails are now made by machine from wire.

Nail strip   Tapered strip of iron (obs) or steel specially rolled for making cut nails.

Narrow end up   *See* Wide end up.

Nasmyth hammer   *See* Steam hammer.

Natural gas   In some overseas countries natural gas is used extensively as a fuel in iron and steelworks. Its use is developing in Britain.

Nature, come to (obs)   When the carbon was burnt out of the iron in a puddling furnace the metal was said to have come to nature or come to grain.

'Neal   *See* Anneal (2).

Neck   The parallel part of a roll which fits into the roll bearing; there is one at each end of the roll.

Necking   In forging, making a groove in a bar, billet, or ingot to mark the beginning of a part to be forged down to smaller size.

Necking tools (spring necking tools)   A pair of tools not unlike fullers (qv) but joined by a long hairpin-type spring at the ends and used with power hammers.

Needled steel   Steel to which a needling agent containing boron has been added. It is claimed to improve the hardenability of the steel.

Negative camber   Some plain work rolls, when used in a four-high mill, have little or practically no spring, and the centre of the barrel is made a few thousandths of an inch less in diameter than the ends, ie the roll is very slightly concave, to allow for expansion caused by heating at work. This is negative camber. Cf. roll camber.

Neilson process (obs)   A process for removing tar and ammonia from blast furnace gases where coal was the fuel.

Neilson stove (obs)   The original form of hot-blast stove (qv).

Nesh   *See* Red short.

Nest of ingots   The group of ingots used with a runner in uphill teeming. *See* Teeming.

Neutral flame   *See* Flame.

Neutral refractories   Refractories such as chrome-magnesite or straight chrome, which are neither acid nor basic. Cf. acid refractories, basic refractories.

Nick and bend test (obs)   A test for wrought iron. A nick was cut on one side of the test piece with a chisel and the nick was opened by bending the test piece back, usually by a succession of light hammer blows. The opened-up nick had to show a fibrous structure and be free from dirt or large pieces of slag.

Nickel   An element, symbol Ni, used in making heat- and corrosion-resisting alloys and, with chromium, for stainless steels.

Nickel steel   An alloy of steel and nickel, especially for heat- and corrosion-resistance. The exact compositions vary.

Nicking   Marking, by means of a nick, an ingot or bar to show the point to which forging is to be done.

Night turn   *See* Turn.

Nipper (obs)   A Sheffield colloquialism for a cellar lad (qv).

Nital   An etching agent used to show up the structure of a steel specimen. It is a solution of 1.5% by volume of nitric acid in methyl or ethyl alcohol.

Nitriding   A case-hardening process. Special types of steel are heated in contact with a source of nitrogen for up to 90 hours. A very thin but extremely hard case is produced on the steel.

Nitrogen   A gaseous element, symbol N, present in air. If picked up by steel during manufacture it can be harmful, but because it will form nitrides, which are very hard, it is sometimes used for case hardening. It can also be introduced beneficially to some special steels.

No 1, 2, 3, 4 or 5 iron   *See* Number one iron.

Nobble (obs) (knobble)   A single puddled ball shingled into a bloom. Two balls welded together were called a pair and the operation was pairing.

Nobbler (obs)   *See* Shingler.

Nobbling (obs)   *See* Nobble.

Nodular iron   *See* Spheroidal graphite iron.

Nodulizing   *See* Balling.

Non-deforming steel   *See* Non-shrink steels.

Non-destructive testing   Any method of examining the soundness of steels without destroying the specimen being tested. The methods include magnetic crack detection, ultrasonic, eddy-current, and radiographic examination.

Non-magnetic steels   Steels of the austenitic type, such as 12% manganese and 18/8 chromium/nickel stainless which are effectively non-magnetic under all normal conditions.

Non-shrink steel (non-deforming steels)   Special alloy steels which do not easily go out of shape or deform when heated. They are used especially for engineers' precision tools and gauges.

Non-slip plates   *See* Anti-slip plates.

Normalize   A form of heat treatment for steels which eliminates internal stresses and improves the mechanical properties.

Nose (1)   The inner end of a tuyere.

Nose (2)   A mass of solidified metal on the inner end of a tuyere.

Nose (3)   The narrow top end of a Bessemer or oxygen converter.

Nose helve (obs)   *See* Helve.

Notch, cinder or slag   *See* Cinder notch.

Notched bar method   A sequence of passes for rolling angles. Starting with a square piece, the rolls first make a notch and then progressively increase it, so forming an angle shape. Cf. butterfly pass, flat and edging pass.

No-twist mill   A name used in USA and Britain for the modern form of rod-finishing mill based on the original Bedson design.

Nozzle

It has alternate vertical and horizontal stands and the piece is thus rolled on opposite sides alternately without the need for twist guides as on a conventional continuous rod-finishing mill.

Nozzle   *See* Stopper.

Number one (two etc) (obs)   A method formerly used for grading pig irons. In general there were four or five grades from 1 to 4 or 5, and mottled iron, but practice varied from district to district and between makers. Some went as far as No 6. Broadly speaking as the number went up the carbon content went down. No 1 would be a grey foundry iron, No 4 a forge iron. The grade was estimated by the appearance of the fracture. Today pig iron is sold on chemical analysis.

Nut iron (obs)   Wrought-iron bars rolled specially for making engineers' nuts.

Nut rods (obs)   *See* Slit rods.

Nuts, smithy   *See* Smithy nuts.

O

Obtuse angles   Rolled angles in which the included angle of the two legs is more than a right angle.

OCP process   The French name (oxygène-chaux-pulverisée) for the LDAC process (qv).

Octagons   Steel (and formerly iron) bars, hot rolled or cold drawn to an octagonal cross section.

Odd man (obs)   The member of a crucible steelmaking team who did all the odd jobs including attending to pot annealing, cleaning away slag, and getting the furnace ready.

Oddsides   Semi-permanent moulds of dry sand or plaster of Paris, used for repetition foundry work.

Oddwork   A Black Country term for general smiths' work, eg small shackles, swivels, door furniture — any small forged parts made for the trade.

Offer up   A term common to many trades and used in forging. It means to try a job in position to see if it fits.

Off-grade (off-heat)   Any steel in which the percentage of any element is outside the specification figure.

Off-heat   *See* Off-grade.

Off-iron   A rare term for pig iron which is not to the required specification.

Offset   See Mismatch.

Off the boil (dead off the boil)   Molten steel in a furnace when all the carbon has been taken out.

OG process   A Japanese method of recovering the gases from an LD converter. The mouth of the vessel is covered by a close-fitting hood, with nitrogen seals, and the gases are collected, cooled, and cleaned. They may then be burnt off to waste in a flare stack or stored and used elsewhere in the steelworks for firing furnaces.

Oil cellar *See* Cellar.

Oil-hardening steel A special alloy steel used for engineers' tools. It is hardened by heating it to a specified temperature and plunging it into an oil bath.

Oil injection Oil injection through the tuyeres of a blast furnace helps to reduce coke consumption. It is now very common. Coal powder has also been injected but the process is not yet successful.

Oil injection fitting A method of fixing couplings to rolling mill rods and shafts so that they can be removed and replaced as required when ball or roller bearings are being changed. The coupling is bored out to an interference fit (ie slightly undersize) and will not normally go right on to the shaft. Oil is injected under high pressure between the coupling and the shaft and this expands the coupling so that it can be pushed home. The oil pressure is then released and the coupling contracts tightly on the shaft. To remove it, oil pressure is applied and the coupling is pushed off.

Oil of vitriol (obs) An old name for sulphuric acid — used for pickling (qv).

Oil sands Moulding sand bonded with oil, eg linseed oil. Used especially for cores.

Oliver (pole oliver) (spring oliver) A spring hammer used by blacksmiths, chainmakers, and other forgers of iron and steel. The hammer is pivoted so that it strikes the anvil and is held in the up position by a horizontal wooden pole (of ash usually) connected to it by a link. Also connected to the hammer is a pedal at floor level. Pressure of the foot on the pedal causes the hammer to strike a blow and the spring pole returns it to the rest position. A skilled user could vary the blow easily. Later olivers often have steel coil springs instead of a wooden pole. They can be divided into pole olivers and spring olivers on this basis.

OLP process (oxygène-lance-poudre process) A French variant of the LD process (qv). It is virtually the same as the

LDAC process (qv).

One-way-fired soaking pit   A soaking pit (qv) which is fired from one end only; it has no reversal of fuel and exhaust gas flows and no regenerators. Cf. regenerative soaking pit.

On stream   *See* Stream.

Open annealing   *See* Black annealing (1).

Open-coil annealing   An American process, also used in Britain, for annealing wide strip in coil form. The tightly wound coil as it comes off the mill is re-wound with the laps separated fractionally. This allows the annealing heat to penetrate more evenly. After annealing the coils are re-wound tightly.

Open-die forgings   Forgings made between dies that are either flat or have a simple groove or V shape.

Opener (obs)   *See* Opening.

Open forepart (obs)   *See* Forepart.

Open-front furnace   A reheating furnace especially for slabs, where the roof is supported at the front by a water-cooled beam, leaving the front completely open when the water-cooled doors are lifted. Long slabs can thus be ejected sideways when hot.

Open-hearth furnace (Siemens open-hearth furnace)   A steelmaking furnace, originally designed by Siemens, fired by gas or oil, in which the charge to be melted is held in a refractory-lined bath, while the flames from the burning fuel pass over it. Firing is from opposite ends of the furnace alternately, and the waste gases pass through chequer bricks to give up a large part of the heat. After about 20 minutes of firing in one direction the flow is reversed. Combustion air then enters through the hot chequers and is preheated before mixing with the fuel. Meanwhile, the waste gases heat a second set of chequers, ready for reversal after another 20 min, by which time the hot chequers will have cooled and the cool ones have become hot again. Open-hearth furnaces can range in size from a few tons to several hundred (even, rather rarely, 500) tons charge capacity. The steelmaking reactions are provided by fluxes and the process can be acid (now rare) or basic. The charge can be molten pig iron, cold pig iron, or scrap or a mixture of two or more of these materials.

Open-hearth process   *See* Open-hearth furnace.

Open-hearth steel   Steel produced in an open-hearth furnace.

Opening (obs) (swording)   Separating pack-rolled sheets which had stuck together during hot rolling. A corner of one sheet was pulled away from the pack and a heavy blunt knife or opener was used to cleave the sheets apart. The knife was variously called the sword, the opener, and the hanker knife.

Open pass   *See* Pass.

Open-sand casting   A casting made in a mould which has no top or cope. The mould is made by pressing a pattern into a bed of sand and filling the shaped depression with molten metal.

Open-sand moulding (obs)   Casting direct into a depression formed in sand by pressing a pattern into it. Now virtually obsolete, but some very early castings (eg firebacks) were open-sand moulded. Only simple shapes could be cast.

Open-top furnace (obs)   A blast furnace with the top open to the atmosphere. Generally the gases were allowed to burn to waste, but some furnaces took off some of the gas by offtakes below throat level.
Cf. closed-top furnace.

Open-topped housing   A rolling mill housing which, instead of being cast in one piece, is made with the top part detachable.

Open train   Any group of mill stands in line side by side, driven through the same set of pinions and accessible from both sides is 'in open train'.

Open tuyere   A blast furnace tuyere consisting of two concentric cones open at the back (or larger diameter). Water sprayed in at the top flowed round the tuyere and out at the bottom.

Optical pyrometer (disappearing filament pyrometer)   A hand-held telescope-like instrument through which a hot body is viewed and its temperature measured by adjusting the electric current passing through a filament until it is of the same brightness as the object being viewed, when it seems to disappear. The voltage is measured by a dial which is calibrated directly in temperature readings.
Cf. radiation pyrometer, immersion pyrometer.

Ore   *See* Iron ore.

Ore bridge   A large gantry crane with a grab, spanning the ore stockyard at a blast furnace, and used to recover ore from stock for use. It may also extend over a dock and be used for unloading ore ships.

Oreing down *See* Pigging down.

Ore preparation Any of several methods of treating ore as it is received from the mine or quarry to make it more suitable for smelting. It includes crushing and screening, and some part of the preparation may be done at the mine or quarry before dispatch.

Osmund furnace A kind of bloomery formerly used in Germany and northern Europe.

O Two (O_2) steel (obs) A Bessemer steel of very high grade, produced by using an oxygen-enriched blast.

Outgoing side *See* Entry side.

Oval pass Any pass in a rolling mill producing an oval cross section (usually before rolling to a round). The rolls in this case are known as oval rolls.

Oval rolls *See* Oval pass.

Oven A device for heating, eg a core oven, to lower temperatures than a furnace. Not used extensively in the iron and steel industry. The word should not be considered as interchangeable with furnace.

Overblow Use of too much air in a Bessemer converter, causing burnt steel.

Overdraft *See* Overdraught.

Overdraught (overdraft) If the rolls in a rolling mill are of slightly different sizes and the peripheral speed of the lower one is slightly less, the piece, on exit, will have a tendency to curve upwards. This condition is often produced deliberately, to prevent the piece striking the mill floor heavily and perhaps running wild or cobbling. The reverse condition is underdraft.

Overfill (flash) (choke) A condition in which there is too much metal to be formed by the rolls of a rolling mill and it spreads out at the roll joint line. Cf. underfill.

Overhead charger *See* Charging machine.

Overhung hammer (Enfield type hammer) An overhung type of steam hammer similar in principle to the Rigby (qv) but having fixed guides.

Overlap *See* Lap.

Overpickling The wrong acid concentration or too long a

period in the vat will overpickle or cause the steel surface to be too rough.

Own arisings   *See* Arisings.

Oxidizing flame   *See* Flame.

Oxy-acetylene   Acetylene gas fed with oxygen to produce an intensely hot flame, used for cutting and gouging steel.

Oxy-fuel lance   A burner using oxygen and oil or gas fuel to provide extra heat for rapid melting down of scrap metal in a steel furnace.

Oxygen   The most common of the elements, symbol O. It plays a vital part in iron- and steelmaking, either in its natural form as a constituent of air or in the artificially produced commercially pure (99.5%) form.

Oxygen lance   A water-cooled tube through which oxygen is injected into a furnace, as in LD etc. It is also used for burning out furnace tapholes.

Oxygen steam process   This is really the VLN process (qv).

P

Pack (obs)   A number of steel (formerly iron) sheets placed one on top of the other for rolling.

Pack annealing   *See* Box annealing.

Packing bars   Steel bars, flat or taper, usually rolled from odd croppings, specially for filling vacant spaces between plates in shipbuilding.

Pack rolling (obs) (ply rolling)   Rolling of sheets in packs until the required thickness is obtained, when the pack is opened, or separated into individual sheets. Also called, rarely, ply rolling.

Padded plates   *See* Anti-slip plates.

Paddle (obs) (bar)   A chisel-ended hand tool used at the puddling furnace during melting down, at the boil, and for quartering (qv).

Pair (1) (obs)   Two sheets hot rolled together.

Pair (2) (obs)   *See* Nobble.

Pairing (obs)   *See* Nobble.

Pallet   A detachable tool face, fitted into dovetails on a drop, steam, or other forging hammer tup and anvil. There are two on each hammer, an anvil pallet and a tup pallet.

Palm end   A form of coupling used on the end of a rolling mill roll.

Palm oil   A vegetable oil used in hot-dip tinning to wash off surplus tin from a tinplate as it emerges from the tin pot.

Pan   The fabricated steel box which holds the hearth lining of the open-hearth furnace.

P and O   *See* CRCA, P&O.

Pane (pean or pene)   The end of a hammer head opposite to the actual hammer face.

Pan handle section   *See* Specials.

Panic button   A colloquial name for an emergency push-button built into a control panel to stop a process quickly if an accident occurs or appears likely. It may also be used to make the process safe if there is an accident and it is not possible to stop it quickly. The cobble shears (qv) are controlled by a panic button.

Parting   A powder dusted or sprayed on the face of a foundry mould to prevent the sand from sticking to that of the second part of the mould.

Parting line (1) (parting plane)   The line along a pattern on which the mould is jointed to allow the pattern to be withdrawn. A simple casting may only have a single parting line; complicated castings may have several.

Parting line (2) (die line)   The plane on which two drop forging dies are divided.

Parting plane   *See* Parting line (1).

Part mine (obs)   Another name for cinder pig (qv).

Pass (groove) (hole)   A pair of matching grooves in a pair of rolls, or the passage of the piece through the rolls. A live pass is one in which work is done on the piece; a dead pass is when the piece is returned over the rolls without being rolled. A closed pass is one in which the groove is enclosed by collars; an open pass has no collars. *See also* Leader pass and Strand pass.

Pass design   The design for a series of passes in a pair or several pairs of rolls, to give progressive reduction and shaping of the piece.

Pass line   The line taken by the piece in its passage through the rolls.

Pass reduction   *See* Reduction (2).

Patented wire   *See* Patenting.

Patent flattened (obs)   *See* Flat sheet.

Patenting   A heat treatment process used on steel wire. It involves heating and then cooling it to the proper temperature in a bath of molten salt or molten lead. Wire so treated is called patented.

Patten rods (obs)   *See* Slit rods.

Pattern   Also called, formerly, a model. A wooden, metal, or plastics facsimile of an article required as a casting and used to form the mould cavity, after which it is removed prior to the metal being poured in.

Pattern draft   *See* Draught (3).

Pattern draw   *See* Draught (3).

Pattern-flow moulding   A foundry system in which the patterns circulate on a track passing through the moulding machines. It is applicable to mass production; if the patterns are left undisturbed the system will keep on making moulds of one type. But a pattern change is quick and easy, and it is also possible to take out a pattern and substitute another for short runs, without upsetting the operation of the line. The system is thus extremely flexible while at the same time having all the advantages of mechanization.

Patternmaker's letters   Metal letters and figures which can be fixed to patterns to enable words to be reproduced on a casting.

Pattern plate   A flat metal plate on which one or more foundry patterns can be fixed for use in moulding machines. When pattern plates are used the moulding process is often called plate moulding.

Pay-off reel   *See* Uncoiler.

Pean   *See* Pane.

Pearlitic malleable iron   A newer form of malleable cast iron produced by a modified heat treatment process, from blackheart iron.

Peat   Has been used, usually mixed with charcoal, as a blast furnace fuel, but never on a large scale.

Peel (1)   A flat-ended tool somewhat resembling a spade, used for putting piles or billets into a reheating furnace, or for putting pieces of scrap in an open-hearth furnace.

Peel (2) (obs) (smoother)   Similar to peel (1) but usually smaller and used for placing fettling in a puddling furnace.

Peel (3)   The gripping or work-holding jaws of a forging manipulator.

Peeled bar   Steel bar subjected to a machining operation to remove scale and surface defects before subsequent treatment such as bright drawing.

Peening, shot   *See* Shot peening.

Peepee   *See* Slag notch.

Peg (stopper)   A metal block placed on the anvil of a press or hammer to prevent the tools from closing below the required distance.

Pelletizing   *See* Balling.

Pellets   Fine iron ore formed mechanically into small round balls and fired to harden them. They are used in the blast furnace. *See also* Balling.

Pendulum mill   A strip rolling mill in which two small-diameter rolls move backwards and forwards, one above and the other below, the strip in a pendulum-like motion.

Pene   *See* Pane.

Peppery blister   *See* Pickling patch.

Pernod puddler (obs)   *See* Puddling machine.

Petro forge   A high-energy-rate forging process in which the forging blow is obtained by rapid combustion of a petrol/air mixture, as in an automobile engine.

Phosphating   A surface treatment for iron and steel, applied (1) as pre-treatment for painting, (2) to provide corrosion resistance, (3) to facilitate cold drawing. It is done by immersing the article to be treated in a hot solution of manganese iron phosphate or zinc iron phosphate. There are several proprietary processes basically of this type. Bonderizing is well known.

Phosphoric cast irons   Cast irons containing not more than about 0.7% phosphorus. The latter element renders the iron more fluid when molten and it can be used for thin-section and intricate castings, but it is also rather hard and brittle.

Phosphorus   One of the elements found in iron ores in varying degrees. It has a harmful effect on wrought iron and steel, making them cold short, and must be removed in processing. *See also* Red-short.

Photomicrograph   *See* Micrograph.

Physic (obs)   Various salts and powders were used occasionally in the puddling furnace with the object of improving the purification of the iron.
Schaffhautl's powder (manganese oxide, salt, and clay) and Scheerer's powder (calcium chloride, salt, and soda ash) are examples.

Piano wire   A very high quality steel wire with 0.8 to 0.95%

carbon and of very high tensile strength (130 to 176 tonf/in^2).
It has special acoustic properties — hence its name — but is used
for general high grade purposes. Sometimes used synonymously
with music wire (qv).

Pickling   Treating the surface of iron or steel with acid to
remove scale, rust, and dirt, preparatory to further processing
such as cold rolling, tinning, galvanizing, polishing, etc.

Pickling blister   *See* Pickling patch.

**Pickling patch (peppery blister) (pickling blister) (strawberry
blister)**   Faults in tinplate caused by incorrect pickling.

Piece   A generic term for any piece of iron or steel being rolled.

Pig (pig iron)   The product of the blast furnace, when cast in a
pig bed or nowadays, in a pig-casting machine. It derives its
name from the fact that the channel or runner leading from the
furnace branched out into side channels called sows and then
into smaller channels called pigs.

Pig bed   *See* Pig.

Pig boiling (obs)   *See* Puddling.

Pig breaker   A machine for breaking the pigs off the sows.

Pig-casting machine   *See* Machine cast pig.

Pig iron   *See* Pig.

Pig iron grades (obs)   *See* Number one (two etc).

Pigging down   Adding pig iron to the bath of an open-hearth
furnace to increase the carbon content as required. The reverse
is oreing down, ie adding iron ore to oxidize the carbon.

Pig lifter (obs) (metal carrier)   A labourer who cleared the blast
furnace pig bed after the pigs had cooled.

Pig's pudding (obs)   *See* Horseshoe main.

Pile (1) (obs) (mill pile)   Wrought-iron bars placed one on top of
the other ready for reheating and rerolling. A pile could be
longitudinal with side pieces (box pile) or in layers alternately
at right angles (cross pile) according to the product being made.

Pile (2) (obs) (ball furnace pile)   A pile of wrought iron scrap
ready for reheating in the ball furnace, for hammering and
rolling.

Piler   A machine for stacking sheets or tinplates as they emerge
from the end of a processing line.

Pill (obs)   A small piece of aluminium used sometimes by a
crucible teemer (qv) to kill the steel before teeming.

Pillar of blast (obs)   An alternative name formerly used for blast pressure on a blast furnace.
A pillar of blast of 2 lb meant 2 lbf/in² .

Pin   A short rod fixed to a lug on the side of a mould cope. It locates in an ear, or lug, on the side of the drag and so guides the two parts of the mould together. A cope and drag usually have three pins and ears each, so that the two halves can only be put together the correct way.

Pincers   A blacksmith's tongs formed like a carpenter's pincers.

Pinch rolls   A pair of rolls between which a strip or bar is gripped and pulled through a processing line or part of the line. The rolls provide traction only and do no actual rolling.

Pinion   In a rolling mill one of the two or three gears which, fixed in a pinion stand, provide the drive from the engine or motor to the rolls.

Pinion box (pinion stand)   A gearbox containing the rolling mill pinions.

Pinion stand   *See* Pinion box.

Pipe   A cavity in the centre of an ingot, caused by the shrinkage of the steel as it cools. If the pipe is in the top of the ingot it is a primary pipe; if in the lower part, a secondary pipe.

Pipe smoother   A moulding tool used to smooth the cylindrical parts of moulds.

Pipe stove (obs)   Any blast furnace hot blast stove in which the air to be heated passed through pipes exposed to the heat. They included pistol pipe, Neilson's, Baldwin's and Staffordshire types.

Pistol pipe stove (obs)   A blast furnace hot blast stove in which pipes were shaped like pistols stood mouth downward. The cold air passed up one side of an internal division and down the other, the whole pipe being exposed to the flames of the stove.

Pitch (1)   The distance between the centres of two adjacent peaks (or valleys) on corrugated sheets. The most common is 3 in. The width of a corrugated sheet is stated as X corrugations of Y pitch, eg 10 of 3 in or 10/3.

Pitch (2)   *See* Coal-tar fuels.

Pit coal (obs) (sea coal) (obs)   Terms used in the period of transition from charcoal to coal and coke firing to distinguish the mineral coal from charcoal.

Pit sample   A sample of molten steel taken during teeming (qv) for checking the chemical composition of the metal.

Pitside   *See* Casting pit (2).

Planetary mill   A heavy, single-pass hot-rolling mill for strip in which two large-diameter support or 'back up' rolls are arranged with a number of small work rolls around the circumferences of each support roll. The whole assembly is so mounted that the work rolls, held by side rings or 'cages', are revolved round the support rolls and the metal to be rolled passes between the work rolls. As much as 90% reduction is achieved per pass and the work done is so heavy that the piece actually rises in temperature

Planish   To produce a particularly flat, smooth surface on a rolled or forged product. It can be done with special rolls or by hammering.

Plastic coating   Many steel products are now coated with plastics material for protection or decoration or both. Finished products can be coated by dipping or spraying and sheets are supplied with a plastics film bonded on one or both sides. Example; British Steel Corporation's Stelvetite.

Plasticity   The property of a metal which enables it to be deformed in any direction and to assume the shape given to it. Cf. elasticity. Steel and wrought iron can be both plastic and elastic according to conditions.

Plate   Wide flat rolled iron (obs) or steel. The dividing line between plate and sheet used to be variable and sometimes obscure. It is now generally accepted that over 3 mm ($\frac{1}{8}$ in) is plate; 3 mm and below is sheet, but although 3 mm is recognized as the dividing line there is some disagreement over the actual category into which 3 mm itself falls. Sometimes it is said that sheet is up to and including 3 mm, sometimes that this dimension is the lower limit of plate.

Plate bricks   Horizontal refractory tubes connecting the runner or trumpet and the ingot moulds in uphill teeming. *See* Teeming.

Plate, charging (obs)   *See* Charging plate.

Plate coolers   Hollow copper plates inserted in the bosh brickwork of a blast furnace, with water circulation to prevent overheating of the refractories.

Plated hearth   If the bottom of a pool of molten metal

Plate metal (obs)

freezes in a furnace hearth, the hearth is said to be plated.

Plate metal (obs) *See* Refined iron.

Plate moulding *See* Pattern plate.

Plate pile (obs) *See* Pile (1).

Plate shear A mechanical shear (qv) for cutting plates.

Plate, standing (obs) *See* Standing plate.

Plating (obs) Welding a piece of carbon steel to wrought iron for edge tools. The wrought iron was then formed into the main body of the tool, the carbon steel forming the hardenable cutting edge.

Plating hammer (1) (obs) Any power hammer used for beating out plates or sheets.

Plating hammer (2) (obs) A steam or other power hammer used for welding carbon steel pieces to wrought iron backing for the manufacture of edge tools, eg spades, shovels.

Plough steel wire A high grade steel wire (originally made in iron) for the ropes of steam ploughing engines — hence the name — but the term is now a general one for wire, especially that of high quality, used in wire rope making.

Plug (1) (obs) A refractory piece used to stop the hole in the bottom of a crucible left by moulding the clay.

Plug (2) (bottom) (cheese) (Holley bottom) The detachable piece, containing the tuyeres, at the bottom of a Bessemer converter. Also called a Holley bottom, after the American inventor.

Plug (3) A steel bar, swollen at one end, used for closing the cinder or slag notch in a blast furnace. It is pushed directly into the aperture in the cooler or monkey.

Plug shop A building in a Bessemer steelworks where the detachable bottoms or plugs for the converters were made.

Plumbago Powdered graphite. Formerly used, mixed with refractory clay, for making steelmelting crucibles, and still used for some refractories.

Ply rolling (obs) *See* Pack rolling.

Pneumatic hammer (air hammer) A power hammer working on the principle of a steam hammer but using compressed air produced in an integral compressor. They were of the over-hung cylinder type, popular for jobbing work.

Pneumatic process Any steelmaking process in which a gas

(ie air or oxygen) is blown into molten iron to refine it. It
includes Bessemer and the various oxygen processes, but all are
better specified by their proper names; pneumatic is too vague.

Pohlig charger   *See* Bucket charging.

Poker-hole cover (obs) (stock rod cover) (obs)   A cast-iron ball
with a small hole right through its diameter, resting loosely in a
cup-shaped casting fixed on top of a gas producer. When turned
so that the hole is vertical a rod can be passed through to poke
the fire. When turned so that the hole is horizontal, the cover is
sealed. The same arrangement was used on single-bell blast
furnaces for the stock rods (qv).

Pole oliver   *See* Oliver.

Pollock slag car   An American design of rail-mounted slag ladle.

Polygram casting   A proprietary type of shell moulding process.

Pony ladle   *See* Ladle-to-ladle degassing.

Poole feeder   A machine for feeding sheets, one at a time, to a
hand tin pot, so partly mechanizing it.

Porous plug   A refractory plug in the lining of a ladle, which is
sufficiently porous to allow a gas to be blown through it but will
not allow molten metal to escape. Gas (inert or air) is blown
through to agitate the metal to ensure proper mixing for a
metallurgical reaction. Cf. shaking ladle.

Port   An opening at the end of and over the hearth of an open-
hearth furnace through which the fuel and air enter and the
waste gases leave. In a gas-fired furnace there are gas and air
ports and in an oil-fired furnace the gas port is replaced by one
for oil fuel.

Portable furnace   *See* Bell furnace.

Porter bar (1) (staff)   A heavy steel bar with a hollow end
which is wedged on to an ingot or forging to support it and
enable it to be turned under the hammer. It is usually longer
than the piece being forged and counter-balances it during
forging. The end of the porter bar may just be hollowed out
to receive the forging or it may be equipped with a chuck to
hold it. Sometimes the hollowed-out end is called the chuck.

Porter bar (2)   A heavy steel bar used to carry and balance
sheet and plate mill rolls as they are taken out of or replaced in,
the mill stand.

Pot (1) (obs)   *See* Crucible (1).

Pot (2) (obs)   *See* Cinder fall.

Pot (3) (obs)   *See* Chest.

Pot annealing   *See* Box annealing.

Pot barrow (obs)   A pair of tongs mounted on wheels and used to transport a crucible of molten steel from the furnace to a casting or teeming point when a large ingot was being made.

Pot clay   A name given, particularly in Sheffield, to a particular natural fireclay originally used for making crucibles and later for other steelworks refractory articles such as stoppers, dozzles, nozzles, etc.

Pot furnace (obs) (pot hole) (hole)   A floor or pot type furnace for crucible steel making. A typical one in Sheffield held two pots and was called a two-pot hole.

Pot hole (obs)   *See* Pot furnace.

Pot house (obs)   The building in which pots or crucibles for steel making were produced.

Pot lid (obs)   A clay lid placed on crucibles when they were in the furnace, to keep out foreign matter.

Pot metal (obs)   An old term for any cast iron suitable for making domestic pots and other holloware.

Pot mould (obs)   A mould used for making steelmelting crucibles or pots.

Pot steel (obs)   Another name for crucible steel (qv).

Poured short (short poured)   A casting which is incomplete because insufficient metal has been poured into the mould.

Pouring basin (runner basin) (runner bush)   A cavity in the top of a mould leading to the runner or gate.

Pouring reel   A reel (qv) into which the bar or rod enters from above via a guide tube; the action looks as if the rod is being poured into the rotating reel.

Powder metallurgy   *See* Sinter (2).

Powder strip   Strip made by pouring iron powder between two rolls arranged side by side, and thus consolidated and rolled into a flat shape.

Power reels   Strip coilers used for pulling strip through a rolling mill, such as a Steckel mill, which is not power driven.

Precision casting (1) (v)   *See* Investment casting.

Precision casting (2) (n)   A casting made to very close dimensional limits, especially by the investment process.

Preform (1) (n)   A continuously cast product with a cross
sectional shape approximately to that of the required finished
shape. The so-called 'dog-bone' is an example. This is a
continuously cast section made for rolling beams. A slice cut off
it looks rather like the conventional drawing of a dog's bone.

Preform (2) (n)   *See* Use.

Pre-heating   A preliminary heating of metal which would be
damaged if raised too quickly to the required temperature. It
may be, and often is, done in a separate zone in the furnace
which does the final heating.

Preparing stamp   A stamping hammer used in conjunction with
a finishing stamp. The piece is partly formed in the first hammer
and finished in the second.

Press (1)   A mechanical or hydraulic machine used for forging,
instead of a hammer. Some presses are very large and exert
forces of several thousand tons.

Press (2)   Similar to press (1) but used for extruding.

Press forging   Forging under a hydraulic or mechanical press,
as distinct from under a hammer.

Press, forging   *See* Forging press.

Presspun process   A combination of pressing and spinning,
used to produce dished ends, as for pressure vessels, and for
flanging.

Pressure casting   *See* Pressure pouring.

Pressure pouring (pressure casting)   A method of casting steel
invented in America. It uses compressed air to force molten
steel up out of a ladle into a graphite mould. Originally for
railway wagon wheels, now used for slabs and other semis
(qv) for rolling. Produces good surface finish and close
tolerances.

Prestressed mill   A modern type of rolling mill stand in which
the rolls are carried in heavy chocks held together by bolts to
which a predetermined load or stress is applied. All mills
stretch under rolling load to some extent; at low loads the
stretch is non-linear or variable and it is difficult or impossible
to keep the roll setting very accurate, which is important
when rolling to close tolerances. At higher loads the extension
becomes linear or at a constant rate. The non-linear extension
is known as the mush zone. A prestressed mill is stretched

beyond the mush zone before it takes the rolling load and rolling accuracy is more easily achieved.

Priler (obs)   *See* Pryler.

Priles (obs)   Sheets which have been hot rolled in a pack of three.

Primary mill   *See* Cogging mill

Primary pipe   *See* Pipe.

Primes   Sheets, especially tinplates, of the highest quality. Cf wasters, menders.

Print   *See* Core print.

Pritchell   A small tool set in the pritchell hole of a blacksmith's anvil and used for punching small holes, eg in horseshoes.

Pritchell hole   *See* Anvil (1).

Producer gas   *See* Gas producer.

Profile   *See* Section.

Protective atmosphere   *See* Controlled atmosphere.

Pryler (obs) (priler)   One of the team at a hand sheet mill (especially in the Black Country). He removed the scale from the hot sheet bar, returned the packs for reheating, and assisted generally. The term was sometimes used as synonymous with scaler or scale breaker.

Puddled ball rolls (obs)   An uncommon name for the forge train (qv).

Puddled bar (obs)   *See* Muck bar.

Puddled iron (obs)   *See* Wrought iron.

Puddled steel (obs)   Steel made from pig iron by decarburizing it in a puddling furnace. It was probably made by decarburizing the iron completely and then recarburizing it to the percentage required. But it may have been done by stopping the decarburization at the appropriate point.

Puddler (obs)   One of two men who worked a puddling furnace. He was in charge; the other was the underhand.

Puddler's ball (obs)   *See* Ball.

Puddler's bar (obs)   *See* Muck bar.

Puddler's candles (obs)   The little blue jets of carbon monoxide flame on the top of the iron during the boil in a puddling furnace.

Puddler's mine   A hematite iron ore used as fettling in the puddling process.

Puddling (obs)   One of two processes for making wrought from cast iron. The older, Cort's, was also known as dry puddling and the newer, Hall's, as pig boiling or wet puddling.

Puddling machine (obs)   Many attempts were made to produce a machine which would puddle iron. None was a success. Danks's mechanical puddler, based on a rotating refractory-lined drum came nearest to success. Clough's was another type, used, but not extensively. Pernod, Gidlow, and Jones were other types.

Pugh type ladle   A form of torpedo ladle (qv).

Pull   *See* Tear.

Pulled strip   Strip of varying width, the result of excessive tension during rolling.

Puller-out (obs)   The man who pulled or lifted the crucibles from the furnace ready for teeming (qv). He also charged the pot and generally supervised the melting.

Pull-over gear   A mechanism used to transfer sideways (or pull over) a piece between two mill stands arranged side by side in line.

Pull-over mill (drag-over mill)   Any two-high mill in which the piece is rolled in one direction only and has to be passed back over the top roll for the next pass.

Pulls   *See* Shrinkage.

Pulpit   A control cabin for a rolling mill or other processing machinery.

Punch   A steel tool used in forging for making a hole in metal being forged. It is used in conjunction with a die, which has a hole in it of the same shape as the punch.

Pup (soap) (closer)   A refractory brick of the same length as a standard square (qv) but thinner or narrower, or both.

Purge   To sweep a furnace or other chamber with an inert gas (usually argon or nitrogen) prior to operation.

Purge gas   Inert gas, usually argon or nitrogen, bubbled through molten steel during degassing, to stir it thoroughly.

Purple ore   *See* Blue billy.

Pusher (1)   *See* Coke pusher.

Pusher (2)   A mechanical ram for pushing pieces into or out of a furnace, or for doing both.

Pusher furnace   A reheating furnace, especially for billets, in which the charge is pushed through mechanically or hydraulically. As a cold billet is pushed in, it pushes a hot one out at the discharge end.

Putter-on-the-line (obs)   In some sheet and light plate mills the shearer had two assistants; the putter-on-the-line, who saw that the sheet was positioned as marked relative to the shear blades; and the runner-round, who took the weight of the back of the sheet.

Pyrometer   An instrument (of many types) for determining, indicating, and sometimes recording temperature.

Q

Quantometer   *See* Spectrographic analysis.

Quantovac   *See* Spectrographic analysis.

Quartering (obs)   Dividing the pasty iron in a puddling furnace, after it had come to nature, into four roughly equal parts with the paddle, ready for balling up and removal.

Quenching   Rapid cooling, part of a heat treatment cycle, eg as in hardening. The cooling medium can be water, one of several types of oil, or a blast of cold air, according to the steel specification.

Quenching car   A mechanically propelled car running on rails between the discharge side of a coke oven and the water-quenching tower, to carry the coke as discharged, for quenching.

R

Rabble (1) (obs)   A hook-ended tool used at a puddling furnace for rabbling during the boil and for balling up.

Rabble (2)   A long steel bar with a flat end used in fettling and maintaining the hearth of an open-hearth furnace.

Rabbling (1) (obs)   *See* Rabble (1).

Rabbling (2)   Stirring a bath of molten steel.

Rabbling (3)   Clearing out holes in a furnace bottom or the sides of the bath during fettling.

Race (obs)   The path taken by the ball from the puddling furnace to the hammer.

Radiant-tube furnace   A heat treatment furnace in which the heat is provided by heat-resisting metal tubes projecting into the furnace space. Fuel (usually gas) burnt in the tubes makes them red hot and the heat is radiated to the charge.

Radiation pyrometer   A pyrometer for use where it is not possible to place thermocouples, eg 'looking' into a furnace. Heat and light from the furnace are focused into a small water-cooled tube, at the bottom of which is a thermocouple, connected to a recording instrument in the normal way. Cf. immersion pyrometer, optical pyrometer.

Radiography   Non-destructive testing of steel for internal soundness by means of a beam of X-rays (produced electrically) or gamma rays (from a radioactive source).

Ragging   Small grooves, cut into the surface of a roll parallel with its centre-line, to help the roll to bite into the piece being rolled. Ragging is only used on roughing rolls; the rag marks on

Rags (obs)   the piece are rolled out during the finishing passes.

Rags (obs)   Wet sacking worn round the legs of a puller out (qv) or teemer (qv) at a crucible furnace, as protection from the heat.

Rail steel   Steel made specially for rolling into railway rails. There are several specifications. A common one contains manganese for toughness and wear resistance. Rails are rolled to customers' own sections or to British Standards and normally in lengths of 60 ft.

Rammer   A tool used for ramming the sand into moulds. It takes various forms on the actual ramming end, such as flat, peg, or ball ended. It is sometimes power operated.

Ram's horn test (obs)   A test formerly applied to wrought iron. A piece of iron had a hole punched into it equal in diameter to $\frac{1}{3}$ of the bar diameter or width and $1\frac{1}{2}$ times the diameter or width from the end. This hole was then drifted out to $1\frac{1}{4}$ times the diameter or width. The end of the bar was then slit up to the hole and the two ends were turned back. All this had to be done without fracture of the metal if the bar was to be accepted.

Randupson process   A foundry moulding process using a mixture of sand, cement, and water for making the moulds.

Rapping   Shaking or jarring a pattern before withdrawing it from the mould.

Rapping bar (rapping iron)   A metal (usually steel) bar used to rap a pattern.

Rapping iron   *See* Rapping bar.

Rapping plate   A metal plate fixed to a wooden pattern and having a hole to receive the rapping bar, which can thus be used without damaging the pattern.

Rastrick boiler (obs)   The earliest type of waste-heat boiler, once used extensively on puddling and mill furnaces.

Rat   A lump on the surface of a casting caused by a piece of sand sticking to the pattern and coming away with it.

Ratchet and ratchet wheel   A lever and slotted wheel fixed to the top of a housing screw, enabling the screw to be raised or lowered by a series of short angular movements of the lever.

Rathole   Small, deep cavities in a steelworks ladle refractories, caused by attack and wear by the molten metal.

Rat tails   A fault in thin flat castings, taking the form of long, narrow furrows on the surface.

Rattler   An American term for a rumbling barrel (qv).

Rattler stars   An American term for rumbling stars (qv).

Raw coal   Mineral coal used in its natural form.

Raw gas (crude gas)   Blast furnace, coke oven, or other furnace gas just as it is taken from the furnace, and before it is cleaned in any way. After cleaning, even partly, it is called clean gas. Gas is not now burnt in stoves or boilers until it has been at least partly cleaned.

Ready iron (obs) (young iron) (obs)   Rare terms for iron which has just come to nature in the puddling furnace and is ready for balling up.

Rebondimeter   *See* Scleroscope.

Recarburization   Adding carbonaceous material (eg anthracite coal) to molten iron which has been decarburized, to bring the carbon level to the required percentage.

Receiver (forehearth)   A refractory-lined reservoir for molten metal in front of a cupola. The term is sometimes applied to a mixer (qv).

Reclamation   *See* Sand reclamation.

Recuperative soaking pit   A soaking pit which is fired usually from the centre and the waste gases pass through recuperators, which heat the combustion air. Cf. one-way fired soaking pit and regenerative soaking pit.

Recuperator   A means of preheating combustion air for a furnace, by passing the products of combustion through heat-resisting metal pipes and passing the incoming cold air round them, or vice versa. Recuperation is essentially a continuous one-way-flow process and is distinct from regeneration (qv).

Red-hard steels   Another name for high-speed steels (qv).

Red hematite   *See* Hematite.

Red sear   *See* Red short.

Red short (hot short) (red sear) (nesh)   A fault in wrought iron caused by excess of sulphur, which causes the iron to be brittle when hot and thus difficult to work. If an excess of phosphorus is present the iron is brittle when cold, or cold short. These terms are becoming obsolete but are still sometimes applied to steel. Also called, less commonly, red sear and nesh.

Reducing flame   *See* Flame.

Reduction (1)   The chemical reverse of oxidation; a process by which the state of oxidation is decreased. It is the basis of all processes for the manufacture of iron from the natural oxide ores.

Reduction (2)   The amount by which a material is reduced in size by rolling. If the amount referred to is that effected by a single rolling pass it is called pass reduction; if it refers to a complete rolling sequence it is total reduction. It is usually stated as a percentage.

Red-work (obs)   The red brick or non-firebrick structure of a brick-built blast furnace. The lining or firebrick part was the white work.

Reek (obs)   To prepare or dress the inside of a crucible steel ingot mould by resting it on a rack over burning tar.

Reel   A machine for coiling bars and rods as they leave the rolling mill.

Reeled bars   *See* Reeling machine.

Reeling machine   A machine which straightens round steel bars by passing them between specially shaped rollers which induce reverse bending. Bars so straightened are called reeled bars.

Refined iron (1) (plate metal) (finer's metal) (metal) (fine metal)
A low-silicon, low-carbon, white cast iron specially prepared in the refinery for use in the dry puddling process.

Refined iron (2) (remelted iron)   Pig iron remelted to a particular specification and recast to pig.

Refinery (obs) (dandy) (running-out fire)   A hearth in which pig iron was heated, usually with charcoal but sometimes coke and partly decarburized to prepare it for dry puddling. Also called a running-out fire. Cf. finery.

Refining   The process of removing impurities from steel by means of slags or by remelting in a special furnace. The final stage of steelmaking is refining. Cf. melting down.

Refractory   A ceramic material which will resist great heat and is therefore suitable for lining furnaces. Fireclay, dolomite, magnesite, and silica are examples.

Regenerative furnace   A reverberatory heating or melting furnace in which the combustion air and gas (if used) are preheated by passing them through regenerators. An open-hearth furnace is an example. If it is oil fired only the air is preheated.

Regenerative soaking pit   A soaking pit fitted with regenerators like an open-hearth furnace. The fuel/waste gas flow is reversed at intervals as in the o-h furnace. Cf. one-way fired soaking pit.

Regenerator   A chamber filled with a honeycomb arrangement of refractory bricks. Hot waste furnace gases are passed through the chambers until the bricks (or chequers) have absorbed sufficient heat, when the hot gases are diverted to another regenerator and air is turned into the first chamber to absorb heat from the chequers. This process of reversal continues at regular intervals and regenerators are therefore intermittent in operation as distinct from recuperators (qv), which are continuous. Regenerators are used on open-hearth furnaces and a Cowper stove is only a regenerator.

Regulator (obs)   A vessel, sometimes with a piston, sometimes without, and sometimes with a water cistern, in the pipe line between the blast furnace and the blowing engine, used to equalize the blast from a reciprocating steam engine. Regulators with a water cistern were called wet regulators, those without, dry.

Reheating   Heating steel from ambient temperature in one of several kinds of furnace to a temperature suitable for rolling.

Reheating furnace   Any one of several types of furnace used solely for reheating (qv).

Reins   The handles of a pair of tongs, as distinct from the jaws or chaps which grip the object being held.

Reladling   Tapping a portion of the charge of an open-hearth furnace into a ladle and pouring it back into the furnace again to obtain as good a mixing of the metal as possible.

Remelted iron   *See* Refined iron (2).

Repeater   *See* Looping mill.

Reroller   *See* Rerolling mill.

Rerolling mill (reroller)   A mill having no iron or steel producing facilities but working solely on semi-finished products supplied by other works.

Residuals   Small quantities of elements remaining in steel after refining. The acceptability or otherwise of residuals depends on the use to which the steel is to be put.

Resistance furnace   A furnace heated by electrical resistances.

Rest bar   *See* Cramp bar.

Returns   *See* Arisings.

Reverberatory furnace   A furnace in which the fuel and metal are separated. Flames from the burning fuel are drawn over the metal by chimney draught and the roof, which slopes downwards, causes the heat to be reflected down or reverberated on to the metal. Puddling, mill, and air furnaces are examples.

Reversal (reversing)   The exhaust gas and fuel flows on an open-hearth furnace have to be reversed about every 15/20 min to reheat the regenerative brickwork. This is done through reversing valves and may be under manual control or automatic. The same principle is applied to the gas and air flows on the blast furnace hot-blast stoves.

Reversing mill   Any rolling mill in which the direction of rotation of the rolls can be reversed at will. Heavy primary mills for bloom and slab rolling are the most common, but others, including some cold rolling mills, are also made to reverse.

Reversing valve   *See* Reversal.

RH process   *See* Ruhrstahl-Heraeus process.

Riffles   *See* Cockles.

Rigby hammer (clear space hammer)   A steam hammer with a single column and the steam cylinder and tup overhung to one side. The tup was guided by the piston rod; there were no separate guides. Cf. Enfield type.
Formerly popular for general purpose forging; the overhung construction left a good clear space for manoeuvring the work.

Rigidized steel (textured steel)   Steel, usually stainless, in thin sheet form on which an embossed pattern has been rolled. The pattern both stiffens the sheet and gives it a decorative appearance.

Rimming steel   *See* Effervescing steel.

Ring and wedge   If a bar is jammed, an oval ring is placed on it near the outer end and a wedge is inserted between the ring and the bar. Hammering the wedge from behind then frees the bar. Also called in the Black Country a Welshman (obs).

Riser   An opening communicating with the cavity in a mould. It fills with molten metal when the mould is poured and acts as a feeder to allow for contraction as the metal cools. It also allows waste gases to escape and, by filling to the top, indicates that the mould is properly filled.

RMS   *See* Mushet's steel.

RN process   An American direct reduction process carried out in a rotating kiln. The name derives from the joint developers – Republic Steel and National Lead Co.

Road   In iron- and steelworks the railway tracks are usually called roads and when close to furnaces are identified by their purpose, eg hot metal road, slag road, charge road, ingot road, etc.

Roak   *See* Roke.

Roast   *See* Calcine.

Robert converter (obs)   A French variation of the Bessemer converter in which the tuyeres were set at a tangent to and below the surface of the molten metal. It was not widely adopted and was superseded by the Tropenas converter (qv).

Robert Mushet's special steel   *See* Mushet's steel.

Robert Mushet's steel   *See* Mushet's steel.

Rocker (1)   A defect in a hot-rolled steel sheet caused by uneven heating before rolling. The sheet elongates unequally, the colder side coming out shortest. Cf. banana-ing.

Rocker (2)   The lever which operates the bellows for a blacksmith's hearth.

Rod   *See* Bar (2).

Rod mill   A rolling mill specially designed for rolling rods.

Roke (roak)   A defect on the surface of a rolled steel bar. It originates from a blowhole which has not welded up during rolling and shows as a fissure elongated in the direction of rolling and only partly closed.

Roll   The cylindrical member which actually forms the metal, in conjunction with one or more other rolls, in a rolling mill. Cf. roller.

Roll bite   *See* Bite.

Roll camber (barrelling)   A small, constant increase in diameter from the ends to the middle (or barrel-shaping) of a sheet or plate roll. It compensates for bending of the roll under load. Cf. negative camber.

Roll-change rig   A mechanical device for changing the rolls of a

rolling mill rapidly, especially sheet and plate mills. It usually
consists of a carriage on a short rail track which can be traversed
right to the side of the mill. The rolls are then withdrawn
through the housing window on to the carriage and moved out
to be taken away by overhead crane.

Rolled-in scale   Scale which has been rolled into the surface of
the metal during rolling. It causes defects on the surface.

Roller (1)   Any cylindrical, rotating component in a rolling mill
or other steelworks plant which is merely for guiding and/or
transferring metal. It does not alter the shape of the metal and
should not be confused with a roll (qv).

Roller (2) (rollerman)   The leader of a team of men at a rolling
mill. Also called a rollerman, especially in Wales.

Roller-hearth furnace   A continuous heat-treatment furnace of
which the hearth consists of power-driven rollers, to carry the
charge through at a preset speed.

Roller leveller   *See* Leveller.

Rollerman   *See* Roller (2).

Roller straightener   A machine with a series of staggered rollers
which induce reversed bending on a bar or section and straighten
it.

Roller table (roll table) (rare)   A line of rollers used for
transferring pieces between stages in rolling. The rollers are
usually set in a straight line with their axes at right angles to the
line of the mills. But they can be set at an angle, when they will
transfer the piece sideways, as in a cross-country mill (qv); the
table is then called a skew roller table. The rollers can be driven
by the passage of the piece or they can be power driven. If
powered they are called live rollers and the table a live roller
table.

Roller twist guide   *See* Twist guide.

Roll force   The total force exerted by the rolling mill rolls in
deforming the piece being rolled. Cf. roll-separating force.

Roll-forging machine   *See* Gap mill.

Roll forming   Forming steel sheet or strip by passing it between
shaped rolls, each pair of which bends the metal progressively
until at the end the required form is obtained. Corrugations are
the simplest example of roll forming (though they can also be
formed in power presses) but much more complicated shapes

are possible. The forming can be done hot or, more usually, cold.

Roll gap   The space between the rolls when there is no piece being rolled.

Roll housings   *See* Housings.

Rolling (1)   Shaping steel by passing it between appropriately shaped rotating rolls.

Rolling (2)   Preparing a cylindrical use (qv) by rounding stock between swages.

Rolling margin (tolerance)   An amount over or under the theoretical weight or dimensions allowed on rolled products. It varies according to the product and is specified in the appropriate BSS. Where there is no BSS the margin is a matter for agreement between buyer and seller.
Cf. exact length.

Rolling mill   Any one of many kinds of machinery for shaping metal, hot or cold. *See* two-high, three-high, four-high, reversing, pull-over, universal, planetary, jobbing, merchant, Lauth, double-duo, cogging, continuous, looping, hand, cluster, Sendzimir.

Rolling mill size   *See* Size of rolling mills.

Roll life   The tonnage of steel rolled by a roll before it is finally beyond repair and is scrapped.

Roll mark   *See* Roll pick-up.

Roll neck   *See* Neck.

Roll-over machine (turn-over machine)   A foundry moulding machine which, when the mould is made, rolls over to enable the pattern to be withdrawn.

Roll pick-up (roll mark)   Defects on a rolled piece caused by a damaged roll, or by foreign matter picked up by the roll and rolled in.

Roll scale   Oxide scale formed during rolling. It cracks and falls off to some extent, but to prevent damage to the piece (especially plates and sheets) it is often blown off by high-pressure steam or water.

Roll-separating force   The force exerted by the resistance to deformation of the piece being rolled.
Cf. roll force.

Roll spring   The total deformation of rolls, housings, and screws,

Roll table (rare)
    caused by the roll-separating force.

Roll table (rare)   Properly, roller table (qv).

Roll torque   The total turning force which must be applied to the rolls to overcome the resistance of the piece to being deformed by rolling.

Roll wobbler   *See* Wobbler.

Rotary furnace   *See* Rotary-hearth furnace.

Rotary-hearth furnace (rotary furnace)   A reheating furnace, circular in plan and having a rotating circular, refractory-covered hearth. Cold steel to be heated is passed through a door in the furnace shell on to the hearth and makes an almost complete circle to be taken out, heated, through a second door near the first.

Rotary shear   A machine with rotating disc-like blades set at the required width, for edge-trimming sheet or strip.

Rotor (Graeff rotor)   A German oxygen steelmaking process in which molten iron is refined in a large, rotating, cylindrical vessel, oxygen, blown in through water-cooled lances being the refining medium. Much less common than the other oxygen processes, eg LD.

Rotten steel   Steel made very badly, or seriously overheated so that when forged or rolled it cracks badly or even disintegrates.

Roughing cinder (obs)   *See* Roughing slag.

Roughing hole (obs)   A sand-lined depression near the cinder fall (qv) of a blast furnace, in which the roughing slag collected and cooled.

Roughing Mill (roughing rolls) (bolting rolls)   A rolling mill used for preliminary rolling; generally at the stage between primary rolling and finishing.

Roughing pass   Generally the first pass through a rolling mill.

Roughing rolls   *See* Roughing mill.

Roughing slag (obs) (roughing cinder)   The slag which followed the iron from the taphole of a blast furnace after tapping.

Roughing stand   Sometimes the first stand in a mill but there may be more than one in a tandem mill; if so they will be numbered.

Round (1)   *See* Bar (2).

Round (2) (obs)   A charge put into a crucible and melted.

Ordinary clay crucibles would not take more than about three; they would then be scrapped.

Round-backed angles   Rolled angles with the back as well as the root having a radius.

Round of charges   The number of skip (or formerly, barrow) loads of ore and coke deposited on a blast furnace bell before it is lowered.

RSA   Rolled steel angle. *See* Angle.

RSC   Rolled steel channel. *See* Channel.

RSJ   Rolled steel joist. *See* Joist.

Rubio   A Spanish iron ore, of about 50% Fe content, formerly imported extensively into Britain.

Ruhrstahl-Heraeus process (RH process) (Top degassing)   A German steel degassing process. A ladle of molten steel is placed beneath a vacuum chamber which has two refractory-lined tubes projecting below it. The legs pass below the surface of the molten steel and the vacuum draws the metal into the chamber. Argon is bubbled up one of the legs to create an upward current. The degassed steel returns to the ladle by the other leg.

Rumbling barrel   A cylindrical vessel in which small castings are placed to be cleaned. Hard iron stars (rumbling or jack stars) are put in with the castings, which they clean by impact as the barrel is rotated slowly.

Rumbling star (jack star)   A piece of hard cast iron, shaped like a star and used in a rumbling barrel for cleaning castings.

Run a melt   To melt down iron in a foundry cupola.

Runner   *See* Pig.

Runner basin   *See* Pouring basin.

Runner bush   *See* Pouring basin.

Runner-round (obs)   *See* Putter-on-the-line.

Running down   A term sometimes applied to melting iron in a cupola. When the charge is fully molten it is run down.

Running heat   *See* White heat.

Running-out fire (obs)   *See* Refinery.

Running stopper   Failure of the stopper in a ladle to close when required. The stream of molten steel continues to run and an emergency is created. The remedy is to lift the ladle over an empty one and let it discharge there.

Run-out table (exit table)   A flat area, sometimes with powered or live rollers, at the finishing end of a rolling mill, on to which the piece runs after the final pass.

Cf. entry table.

Ruptured steel   Steel which has been forged heavily in thin cross sections and torn or ruptured.

Rust   A hydrated oxide of iron $2Fe_2O_3.3H_2O$, formed by iron on exposure to air and moisture.

Rust-resisting steel   *See* Stainless steels.

S

Saddening   Light forging preparatory to reheating ready for further, heavier forging.

Safety lining   A separate lining between the working lining and the wall of a ladle which acts as its name indicates in case of breakdown of the working refractory.

Salamander (1)   Another name for a bear (qv).

Salamander (2)   Sometimes used for the bloom made in a bloomery (qv).

Salt bath   A vessel containing molten salts, used for heating steel for hardening or tempering. Different salts are used according to the temperature required. Typical salts are: for tempering, sodium and potassium nitrate; for hardening, sodium cyanide and barium, calcium, potassium, and sodium chlorides.

Sample passer   Is responsible for deciding, on the basis of samples taken from a steelmaking furnace, when the melt is to specification and the furnace can be tapped.

Sampling spoon   *See* Spoon.

Sandblasting (shotblasting)   Cleaning the surface of iron or steel by blowing sand at it from a nozzle at high speed. The use of natural sand has been discontinued because of the health hazard and iron or steel shot is now used, so the process should be called shotblasting, but the old name lingers on.

Sand casting (1) (v)   Making castings in sand moulds.

Sand casting (2) (n)   Any casting made in a sand mould.

Sand core rolling   A method of producing hollow steel for

Sand cutting

pneumatic drills. A billet was pierced, the hole filled with sand, and the ends closed with plugs. The billet was then hot rolled to finished size, the plugged ends were cut off, and the sand was washed out by water under pressure.

Sand cutting   A term sometimes applied to the preparation of sand, by hand or machine, ready for moulding.

Sanding (obs)   When hand-made crucibles were used they had a small hole in the bottom where the guide pin for the forming plunger passed through. When the pot was placed on the stand (qv) a little sand was thrown into the bottom of the pot, the action being called sanding. The sand 'fritted' (became semi-plastic) in the heat of the furnace, so sealing the hole and firmly cementing pot and stand together. Pots made by machine had no hole and no sanding was necessary.

Sandless pig iron (obs)   *See* Machine-cast pig.

Sand mill   Several kinds of machine are used to prepare foundry sand mixtures by mixing the various ingredients mechanically.

Sand mould   A foundry mould made by any one of several moulding processes using sand, eg green sand, dry sand, (qv).

Sand muller   *See* Muller.

S and O (obs)   *See* Sheared and opened.

Sand reclamation   Processing, usually mechanically, of used foundry sand, to remove dust, metal, and foreign matter and introduce the correct amount of moisture, to prepare it for re-use.

Sandwich rolling   *See* Cladding.

Sap (obs)   The uncarburized centre part of a blister steel bar.

Sash bar   A rolled steel special section made for window and roof glazing.

SAT   *See* Standard area of tinplate.

Sates   *See* Set hammers.

Saw, cold   *See* Hot saw.

Sawdust   *See* Branning.

Saw, hot   *See* Hot saw.

Say ladle   *See* Spoon.

Scab   A fault in the form of a blister or swelling on the surface of an iron or steel casting.

Scaffold   *See* Hanging.

Scale   Oxide of iron formed on the surface of iron or steel by heating.

Scale breaker (1) (obs)   *See* Pryler.

Scale breaker (2)   A series of roll stands, or often a single stand, which is designed to impose heavy rolling pressure to break up the surface scale to allow it to fall or be blown off. Alternatively a scale breaker may be a group of rolls arranged horizontally like a roller leveller (qv).

Scale car   A weighing hopper running on rails and self-propelled. to collect weighed charges of ore and other raw materials for a blast furnace or cupola.

Scale pit (1)   A surface fault in forgings, caused by scale being forced in during forging.

Scale pit (2)   A pit in the floor near a rolling mill stand into which mill scale falls or is washed and is removed for disposal.

Scaler (obs)   *See* Pryler.

Scalping   Machining the top layers of ingots, billets or slabs. Scalping is more commonly done on non-ferrous metals than on steel.

Scarf   To forge the end of a bar to tapered spade form to match a similar scarf for welding two bars together or to weld a chain link. Cf. jump weld.

Scarfing   Burning out surface defects from semi-finished steel prior to final rolling.

Scarr (obs)   A lump of semi-fused iron produced by overheating in the calcining kiln. It was broken up with a heavy iron bar by a man called a scarrer.

Scarrer (obs)   *See* Scarr.

Schaffhautl's powder (obs)   A 'physic' (qv).

Scheerer's powder (obs)   A 'physic' (qv).

Scleroscope (Shore scleroscope) (rebondimeter)   A hardness testing instrument used especially on rolling mill rolls. A small weight is allowed to drop freely down a graduated glass tube and strike the surface being tested. The hardness is measured by the height to which the hammer rebounds.

Scorched ingot   An ingot which is brittle through being teemed too hot.

Scoria   An alternative name for slag, now little used.

Scotch cleaner   A foundry moulder's tool. It is a long flat piece of metal with one end turned at right angles to reach into otherwise inaccessible corners when cleaning up or repairing a mould.

Scotch pig (obs)   Pig iron made in Scotland for light or ornamental castings. It used to be made from local clay ironstone or blackband ores but these are not now worked, and the name is really only of historical interest, though pig iron is still made in Scotland.

Scotch tuyere   *See* Tuyere.

Scouring (obs)   A primitive method of dressing iron ore. The ore was thrown into a stream and left for a time for the water to wash away the lighter impurities, then it was shovelled out. Alternatively it was spread on the ground and water was caused to wash over it by breaching a small dam, made for the purpose.

Scrap ball (1)   *See* Drop ball.

Scrap ball (2) (obs)   A ball of crop ends and selected scrap worked up in a puddling furnace after fettling to glaze the bottom, or in a mill furnace to use the scrap.

Scrap cutter   A man cutting up sheet or other scrap with shears into disposable pieces.

Scraped edge   The edge of a strip damaged by incorrectly adjusted guides.

Scrap end (obs)   *See* Curl end.

Scraper (obs)   A hook-ended hand tool used when tapping cinder from a puddling furnace.

Scraping (obs)   *See* Matching.

Scrap lad (obs)   A boy who worked at a sheet shears, cutting the scrap into small pieces with a small shear, for disposal.

Scrap pack (obs)   *See* Curl end.

Screen bars (grate bars) (wedge bars)   Steel (formerly iron) bars rather like flats but tapering in cross section. Used with the thick end upwards in screens and fireboxes.

Screw, housing   *See* Housing screw.

Screwdown   The housing screws, nuts, levers, spanners (qv), and other gear (including electric or other power equipment when fitted), used for setting the rolls in a rolling mill. It can be remotely controlled.

Screw rods (obs)   *See* Slit rods.

Scrubber   A vertical closed vessel in a furnace gas system, in

which the waste gases ascend through sprays of water.
Dust is washed out of the gas and collects as a sludge at the
bottom of the scrubber.

Scruff (heavy metal)   An iron-tin compound which forms in the
tin pot during hot-dip tinning. It has to be removed from time
to time or the pot would fill with it.

Sea coal (obs)   *See* Pit coal.

Seams   Long shallow surface defects in semi-finished or finish-
rolled steel.

Season   *See* Temper (5).

Seasoning   *See* Weathering.

Secondary pipe   *See* Pipe.

Second-hand melter   *See* First-hand melter.

Seconds   *See* Wasters.

Section (shape) (profile)   Any rolled product which is not a
round, square, or flat. This is British usage. In USA the term is
often shape and in Europe, profile.

Sectional core   A foundry moulding core made in two or more
parts and wired together or stuck together with special adhesive.

Section mill   A rolling mill designed to roll sections such as
joists, channels, tees.

Seger cones (fusible cones)   A range of small cones of known
melting point. To test the temperature of a furnace a cone of
the appropriate melting point is placed in it. At the marked
temperature it melts and collapses.

Segregation   Non-uniform distribution of impurities in steel. It
takes many forms and is influenced by the chemical composition
and the rate of cooling of an ingot.

Selective carburizing (selective hardening)   Introducing carbon
into selected parts only of a piece of steel so that those parts
alone can be case hardened. The parts which are not to be
carburized are copper plated, packed in sand or coated with a
proprietary solution or paste.

Selective hardening   *See* Selective carburizing.

Self-fluxing ores (self-going ores)   Iron ores which contain
naturally enough lime to yield a proper slag in the blast furnace;
consequently no limestone needs to be charged with them.

Self-going ores   *See* Self-fluxing ores.

Semi (n)   Any semi-finished rolled steel or iron not suitable for sale until further rolling has been done. Plural semis.

Semi-continuous mill   A rolling mill in which some of the stands are arranged in tandem for continuous rolling and some are side by side, usually looping.

Semi-killed steel   *See* Balanced steel.

Semi-steel (ferrosteel)   The product of melting mild steel scrap and cast iron together in a cupola. It is a misnomer, for the metal is still cast iron, though of lower carbon content.

Sendzimir mill (Z mill)   A multiple-back-up cluster mill for cold rolling sheet or strip to very precise limits and fine finishes.

Septum valve   A throttling valve in a blast furnace gas offtake system, which maintains high top pressure (qv) automatically.

Set down   A diameter change when producing a forging.

Set hammers (sets) (sates)   Tools of various shapes and sizes used by the blacksmith. They are fixed in handles and used where striking direct with the hammer would be impracticable. Sets are held by the smith and struck with a sledge hammer by the striker. *See also* Chisel.

Sets   *See* Set hammers.

Setting down   Reducing the diameter or cross-sectional area of a forging.

SG iron   *See* Spheroidal graphite iron.

Shadrach (obs)   A ball drawn from the puddling furnace before the iron had come to nature; it would not stand shingling.

Shaft   *See* Stack (1) and (2).

Shaft furnace   A furnace consisting primarily of a vertical shaft, usually circular, in which metal is heated, melted or smelted. The blast furnace and cupola are the commonest examples but a few others are so constructed, eg some continuous annealing furnaces.

Shakeout   *See* Knockout.

Shaker-hearth furnace   A continuous heat treatment furnace for small components in which the complete hearth is given, mechanically, a shaking motion. This causes the components to be shaken along the hearth, so proceeding from the loading to the discharging end. The length of hearth and shaking motion are so designed as to expose the components to the heat of the furnace for the required time, and the heat treatment cycle is completely automatic.

Shaking ladle   A type of ladle, usually closed at the top during use to prevent loss of heat, and given a circular shaking motion. A rotating wave forms on top of the molten metal, breaking it up and accelerating the metallurgical reactions.

Shank   *See* Ladle.

Shape   *See* Section.

Shapes of ingots   Ingots are usually
(a) rectangular (slab ingots)
(b) square (bloom ingots)
(c) polygonal (8 or 12 sided, for forging)
(b) and (c) can be fluted (ie corrugated) to increase the surface area and reduce the tendency to cracking on cooling.

Sharp sand   Sand which does not bond together naturally. Some moulding sands contain the sharp variety.

Shear   A machine for cutting steel bars, plates, sheet, etc by a mechanical cutting action. It can be of many sizes from a small hand machine used by a blacksmith to a power driven machine with a force of hundreds of tons, and can be used for cutting into required lengths or for cropping (qv).

Shear, crocodile   *See* Crocodile shear.

Sheared and opened (obs) (S & O)   The stage in hot pack rolling when the turned-over edges of the pack are sheared off and the individual sheets are separated by opening.

Sheared edge   *See* Edge (2).

Shear end   *See* Curl end.

Shearer (cutter)   A man in charge of a mechanical shear.

Shear, flying   *See* Flying shear.

Shear, guillotine   *See* Guillotine shear.

Shear steel   *See* Cementation.

Sheet (1)   *See* Plate.

Sheet (2) (obs)   In the Black Country sheets from 3 to 20 gauge were called singles, from 21 to 24 G doubles, from 25 to 27 G triples, trebles or lattins, and from 28 G downwards double doubles or bosters. All now obs.

Sheet bar (obs)   A hot rolled, semi-finished iron or steel bar made specially for rolling out to sheet. In Wales often called an iron, even if of steel.

Sheet bar multiple (obs)   A length of flat rolled bar from which a definite number of sheet bars could be cut.

Sheet furnace (obs)   A reheating furnace specially for reheating sheets between rollings.

Sheet piles (interlocking piles)   Flat rolled steel with special edges made to interlock by sliding one within the other. Used by civil engineers for walling, coffer dams, etc, these piles are really much too thick in section to warrant the use of the word sheet in their name but it is generally accepted.

Sheet sizes (obs)   Iron sheets were known as singles, doubles, and trebles or lattins according to their thickness and method of rolling. The actual thicknesses varied but the following are typical: singles, between 4 and 20 BG (0.250 to 0.039 in); doubles, between 20 and 25 BG (0.039 to 0.022 in); trebles, between 25 and 27 BG (0.022 to 0.017 in).

Shell (splash)   A defect on the side of an ingot, caused by metal, splashed during teeming, having solidified and stuck to the mould wall.

Shell core   *See* Shell moulding process.

Shell moulding process   Or Croning process after its German inventor. A moulding process using a mixture of sand and chemical binder. The sand is poured on to a hot metal pattern, where the binder softens and the sand takes up the shape of the pattern. The mould is then baked for a few minutes, when it becomes a firm shell shaped like the pattern. Moulds and cores are made in this way, often in sections and chemically glued together or embedded in sand to hold them tightly in place while the metal is poured.

Sherardizing   A zinc coating process, in which iron or steel articles are heated in contact with powdered zinc.

Shift (1)   A casting fault caused by misalignment of cope and drag.

Shift (2)   *See* Turn.

Shingle (obs)   To hammer a puddled ball to expel the surplus slag and form it into bloom for rolling. Sometimes called nobbling.

Shingler (obs) (nobbler)   The workman at a shingling hammer. If the hammer was steam he was assisted by a hammer driver. Rather rarely called a nobbler.

Shirt   Sometimes used in America as an alternative to the term blast furnace lining.

Shoe   A girder fixed parallel to a rolling mill rolls and used to carry the housings.

Shoe iron   *See* Horseshoe iron.

Shore scleroscope   *See* Scleroscope.

Short poured   *See* Poured short.

Short weight (obs)   Two standards of weight were formerly common in the iron industry — short weight (112 lb = 1 'hundred' or cwt) and long weight, (120 lb = 1 'hundred' or cwt). The ton was of 20 cwt in both cases, so a short ton was 2,240 lb and a long ton 2,400 lb.

Shot (1)   Cast iron or steel formed into pellets by dropping thin streams of molten metal from a height into cold water. It is used for shotblasting.

Shot (2)   *See* Cold shut.

Shotblasting   *See* Sandblasting.

Shot peening (cloudburst hardening)   A method of surface hardening steel by allowing hard steel balls or shot to fall or shower on it. The surface is thus work hardened.

Shotting   Making shot (qv).

Shrinkage (sinks) (draws) (pulls)   Rough cavities in castings caused by incorrect gating and feeding; there is insufficient molten metal as the casting cools.

Shrink rule   *See* Moulder's rule.

Shut (1)   *See* Lap.

Shut (2)   *See* Cold shut.

Shut (3)   A blacksmith's term for making a weld, particularly in a chain link.

Shut plate   *See* Spade.

Side-blown converter   *See* Tropenas converter.

Sideguards   A pair of heavy plate-like devices mounted vertically in a rolling mill roller table and so arranged that they can be brought up mechanically to clamp a piece and move it sideways. They are a form of manipulator (qv), but can only slide the piece sideways; they cannot turn it on edge.

Siderite   A form of iron ore $FeCO_3$. Contains up to 48% Fe.

Siding   Steel sheet for cladding the sides of buildings. It may be ordinary corrugated sheet or specially-formed sections.

Siemens gas   *See* Gas producer.

Siemens-Martin process (Martin-Siemens process — rare)   A

French modification of the Siemens open-hearth process (qv). It was based on the use of cold pig iron and wrought iron scrap together in the furnace. The term was formerly used for any Siemens open-hearth operation.

Siemens open-hearth furnace   *See* Open-hearth furnace.

Siemens producer   *See* Gas producer.

Siemens puddling furnace (obs)   A puddling furnace of orthodox form but fired by producer gas and equipped with regenerators. Not much used.

Silica   Silicon dioxide (SiO_2). Quartz and sand are forms of silica. It is used as an acid refractory.

Silica brick   A refractory brick containing at least 92% silica.

Silicate cotton   *See* Slag wool.

Silky pig iron   Blast furnace iron with a high silicon content and a bright, glazed appearance when fractured.

Sill   The horizontal member on which the door of a furnace rests when closed.

Silver steel   Bright drawn or ground carbon steel (about 0.95 to 1.25% C). It is used by engineers and has no silver in it; the name derives from its appearance.

Single-pass mill   A modern German rolling mill development, based on a Swiss design, in which finish hot shaping of bar stock can be done in a single stand. Current examples actually have two stands and are therefore strictly two-pass, though they generally retain the title single pass.

Singles   Iron (obs) or steel sheets which have been rolled singly (ie not in packs). See sheet sizes.

Single-sweep tin pot   *See* Tin pot.

Sinks   *See* Shrinkage.

Sinter (1) (n & v)   Agglomeration of small pieces of iron ore and dust by heating to provide a uniform blast furnace burden (qv).

Sinter (2) (powder metallurgy)   To make small metal components (including iron), by pressing fine metal powder in a shaped die and heating to form a firm component by diffusion, grain growth, and recrystallization.

Sinter cooler   A large rotating bed at the discharge end of a sinter strand, where cold air is drawn through the red-hot sinter to cool it.

Size of blast furnaces   In Britain the size of a blast furnace is specified in terms of hearth diameter. Some overseas countries use the interior volume as the measure. The use of the bosh diameter is obsolete.

Size of rolling mills   For all except sheet and plate mills the size of a mill is stated in terms of nominal roll diameter. For sheet and plate mills the size is the roll barrel length. The modern practice for sheet and plate mills is to give both diameter and barrel length, diameter being stated first. The custom for British mills has always been to give the size in inches, eg 8 in guide mill, 48 in plate mill.

Sizing   *See* Coining.

Skeleton pattern   A foundry pattern made in the form of a frame. The surface of the mould is finished off by the moulder with hand tools.

Skelp   Strictly, a flat strip, with tapered edges, made specially for making tubes by drawing them through a bell or die at welding temperature. The tapered edges, when the strip was turned round, approached each other parallel, so making a good butt weld. The term is often applied to any strip used for tube making.

Sketching   Marking out steel plates for shearing.

Skew roller table   *See* Roller table.

Skim   To hold back dirt, slag, or scum from the main stream of molten metal. It can be done manually with a hand tool called a skimmer, by a refractory brick or by a refractory-coated metal plate, called a weir or skimmer.

Skimmer   A dam of sand across the blast furnace runner near the taphole, to hold back slag floating on the iron and divert it to the slag runner for disposal.

Skimming rolls   A pair of rolls at the exit end of a sheet galvanizing bath. They draw out the sheets and squeeze off surplus zinc.

Skin pass (temper rolling)   A very light cold rolling pass on steel sheet, done for metallurgical reasons at the end of the rolling and heat treatment sequence.

Skip hoist (charging skip)   A wheeled container, power hauled up a steep incline to charge a blast furnace or cupola. It inverts itself automatically at the top to discharge.

Skull (1) (obs)   Burnt or spoilt wrought iron.

Skull (2)   The shell of metal and slag which builds up on the lining of a steel ladle, in a runner or launder, or remains in a furnace.

Skull cracker or breaker   The man who breaks up skulls for remelting, also a machine for this purpose.

Slab   A semi-finished hot-rolled piece, of flat section, prepared for rolling down to plate or sheet. It is generally more than $1\frac{1}{2}$ in thick and more than twice as wide as it is thick.

Slabbing mill   *See* Cogging mill.

Slabbing pass   In rolling slabs most of the passes are given on the broad, flat sides (slabbing passes), but several times during the rolling sequence the slab is turned up on edge and the edges are rolled (edging passes).

Slab shear   A mechanical shear (qv) for cutting slabs.

Slack blast   If production at a blast furnace is to be halted for a short time (because of plant breakdowns or temporary hold-up of raw material supplies) the blast can be taken off, leaving the furnace to burn slowly. This is known as putting the furnace on slack blast. It should not be confused with banking (qv).

Slag (1) (cinder)   The non-metallic impurities removed from iron ore in a blast furnace during smelting and drawn off in molten form.

Slag (2)   The non-metallic material forming a molten layer on the top of molten steel in a steel furnace. It is made by charging suitable materials and plays an important part in the refining of the steel.

Slag (3)   Loosely applied to any waste material drawn off in molten form.

Slag bank (cinder bank) (cinder hill)   A tip for waste slag, especially from blast furnaces.

Slag bott   A long knob-ended hand tool used for stopping up a blast furnace slag notch.

Slagceram   A British proprietary material made by melting blast furnace slag and casting it into shapes suitable for abrasion-resisting surfaces.

Slagger (teazer)   The man who taps the slag at a blast furnace.

Slagging (flushing)   Tapping the slag from a furnace. At a blast furnace it is often called flushing.

Slag granulation   *See* Granulated slag.

Slag notch (cinder notch) (flushing hole) (monkey) (peepee)
The aperture through which slag is tapped from the blast
furnace.

Slag patch   Slag trapped under the surface of a steel ingot as it
freezes. If it is elongated it is called a slag stringer.

Slag pocket   A chamber built into each end of an open-hearth
furnace below the hearth, to catch slag carried over by the
waste gases and prevent it entering the regenerator chambers.

Slag-short   Steel which is embrittled by slag inclusions.

Slag stringer   *See* Slag patch.

Slag wool   An insulating material made by blowing steam
through a stream of molten blast furnace slag, causing it to
divide and solidify in fine threads. The term mineral wool is
usually taken as synonymous with slag wool, but mineral wool
may, in fact, be made from some other artificial or even
natural stone.

Sledge   A heavy steel plate inserted under the bottom roll in a
rolling mill and drawn out by power to bring out the roll for roll
changing. In some large modern mills both (or all four) rolls can
be drawn out on the sledge through the housing window.

Sleeker (slicker)   One of several types of tool used by a moulder
for smoothing parts of moulds.

Sleeve   A tubular refractory brick. Several are used on the
stopper rod of a ladle to protect it from the molten steel.

Sleeved rolls   Rolling mill rolls in which the body is made of one
material and the working surface of a separate sleeve, shrunk on;
eg an SG iron sleeve on a steel body or arbor.

Slice   A small spade-ended tool used by a blacksmith to control
his fire.

Slicer (ingot slicer) (slicing machine)   A rotating cutting head
which encircles an ingot and cuts slices off for forging.

Slicing machine   *See* Slicer.

Slicker   *See* Sleeker.

Slide gate nozzle   A replacement for the common stopper (qv)
and nozzle on a steelworks ladle. It has a refractory slide working
in refractory guides and can be drawn back to allow a stream of
molten metal to pass through the nozzle in the bottom of the
ladle. This type of valve permits very accurate control of the

metal stream. It can also be used in tundishes (qv), eg for continuous casting, and on certain types of furnace.

Slime   A term borrowed from metal mining. Very fine particles of iron ore are usually called slimes. They are finer than dust and can even be colloidal.

Slip (1)   A condition in which the rolls of a rolling mill fail to grip the piece and it does not move.

Slip (2)   *See* Hanging.

Slipper   A radiused component which allows the palm-end coupling of a rolling mill drive a certain amount of universal motion.

Slit edge   *See* Edge (2).

Slit iron   *See* Slit rods.

Slit rods (obs) (slit iron)   Narrow rods cut from strip or sheet in the slitting mill, mainly for nailmakers but also for making screws, chapes (for buckles), nuts, pattens, and sprigs (small nails), and designated screw rods, chape rods etc.

Slitting machine   The modern, precise, form of the old slitting mill (qv), used for edge trimming of strip and/or for dividing wide strip into narrow strip.

Slitting mill (obs)   A mechanically driven mill of two parts, (1) plain rolls for reducing hammered bar to strip, (2) a pair of rolls with intersecting collars to slit the strip into rods.

Slot oven   An American term used to distinguish the modern form of multiple-chamber coke oven from any of the older type such as the beehive.

S-L process   A direct reduction process using pelleted iron ore with anthracite or coal as the reducing agent. It is carried out in a rotating kiln.

Slug (1)   In drop forging, a piece of bar or billet just big enough to make one forging.

Slug (2)   A special charge introduced into the normal charge of a blast furnace with the object of cleaning accretions from the lower stack and bosh wall. It may consist of coke and scrap or of low-melting point siliceous material such as mill cinder, fluorspar, or Bessemer slag.

Smelt   To reduce an ore to liquid metal. Not to be confused with melt (qv).

Smelter   This term is mainly confined to the non-ferrous

industry, though it could correctly be applied to a blast furnace,
which is a device for smelting iron; its use thus is rare. But the
word has also been misused in the past for 'melter' (qv).

Smiddy coal (obs) (Scot)   Smithy coal. *See* Smithy nuts.

Smith forged   *See* Hand forged.

Smith's composition   *See* Angus Smith's composition.

Smith welding   *See* Forge welding.

Smithy   The workshop and its equipment in which forgings
are made. It can range from a small shop for works maintenance
to a large department making forgings for sale and may or may
not have power hammers and other machinery. Cf. Forge (2).

Smithy char   Coke formed by coking small coal in the smith's
hearth.

Smithy coal   *See* Smithy nuts.

Smithy nuts (obs) (smithy coal)   Bituminous coal, in pieces about
the size of a Brazil nut, used in smiths' hearths.

Smoother (obs)   *See* Peel (2).

Smothering (obs)   A stage in puddling, just before the boil,
when the damper was put down to prevent burning the iron
by giving a smoky, reducing atmosphere.

Snap flask   A moulding box which is hinged at one corner
and fitted with a clip or clips at the opposite one so that it can
be removed from the mould after moulding. After removal of
the snap flask a metal or wooden frame or jacket is placed
round the mould to hold it together.

Snapper   *See* Cinder snapper.

Snatcher   A hook-ended tool used for extracting tuyeres
(tuyere snatcher) or coolers (cooler snatcher) from a blast
furnace.

Snort valve   A valve in the cold-blast main of a blast furnace,
operated by a hand wheel in the cast house (the snort wheel)
to open the main to atmosphere and so take off the blast
without stopping the blowing engine.

Snort wheel   *See* Snort valve.

Snowball heat   *See* White heat.

Soak   To hold a piece of steel at a fixed temperature long
enough for the heat to penetrate or soak uniformly throughout.

Soaker   *See* Soaking pit.

Soaking pit (soaker)   Originally just a refractory-lined pit in which ingots were placed immediately after stripping them from the moulds, to allow the heat from the hotter centre to soak out until the whole ingot was more or less uniformly hot and ready for rolling. Modern soaking pits have an external source of heat (eg gas or oil fuel) and may be regenerative or one-way fired.

Soap (1) *See* Pup.

Soap (2) A lubricant akin to, but different from, domestic soap, used in drawing wire.

Soda ash Sodium carbonate ($NaCO_3$). It is sometimes used for desulphurizing molten iron in a ladle, the sulphur coming off as sodium sulphate.

Soft blast Weak or low pressure blast.

Soft skin A thin surface layer on high-speed or other alloy steels which has become decarburized during heating. It is a fault, which has to be prevented by various means, eg protective atmospheres in furnaces.

Soleplate A specially rolled steel section for cutting up into soleplates, on which flat-bottomed railway rails rest. The sections are generally called soleplates, though strictly they are not until cut into the required lengths.

Solid-bottom cupola (obs) The older form of cupola had a solid bottom, like a blast furnace with an access hole or breast hole at the front. It has been superseded by the drop-bottom cupola (qv).

Sow *See* Pig.

Space bar *See* Z-section.

Spade (checker) (shut plate) A clay-coated iron or steel plate with a ring-ended handle, used to stop the flow of iron in a runner when a row of pig moulds was full.

Spalling (1) (flaking) Breaking off of small pieces from the surface of metal. It can happen to hard steels being worked, or to a rolling mill roll during operation.

Spalling (2) Disintegration of furnace refractories, by pieces breaking off.

Spangle A glitter effect on the surface of a galvanized sheet, caused by the crystals of zinc. The size of the spangle can be controlled during processing; it can be eliminated altogether, leaving a matt surface.

Spanner   The lever on top of a housing screw, used to turn it.

Spar   *See* Fluorspar.

Sparking (spark testing)   Identifying a type of iron or steel by grinding it and observing the colour and nature of the sparks. The difference between wrought iron (dull red sparks) and mild steel (bright sparks) is easy to see, but a skilled operator can go much further, and recognise different carbon contents.

Spark testing   *See* Sparking.

Sparry ore   *See* Spathose ore.

Spathic ore   *See* Spathose ore.

Spathose ore (spathic ore) (sparry ore)   An iron ore of crystallized ferrous carbonate, $FeCO_3$ (ferrous oxide) combined with carbonic acid.

Specials (special sections)   Sections rolled in steel (formerly iron) for special purposes, often to a customer's own specification. Many can be found in old records, the purpose of which is not now known, but a few examples can be given. Boat guard iron, a flat with two rounded edges, was used to protect wooden canal boats; tip iron was used for boot tips; key section was for engineers' (not locksmiths') keys; casement section for glazing; thimble section for the rope trade; bucket or pan handle sections are obvious. Bobbin sections (shaped in cross section roughly like a bobbin) are often used for railing uprights.

Special sections.   *See* Specials.

Special steels   A vague term, formerly used to distinguish alloy steels from carbon steels but sometimes applied to any steel which is considered to have some special qualities.

Spectacle valve   *See* Goggle valve.

Spectograph   *See* Spectographic analysis.

Spectographic analysis   A method of analysing a sample of iron or steel by striking a small electric arc on it and identifying the characteristic spectra emitted by the various elements. Identification can be done by electronic equipment automatically, the analysis of the sample being typed out by the instrument, which is very fast in operation, only a few minutes being needed for 12 or more elements. The Quantovac and Quantometer are proprietary automatic spectographs.

Specular iron ore   Crystallized ferric oxide, with the same composition as hematite.

Spelly wire   Wire which is defective through segregation and likely to fracture during drawing.

Spelter   A name formerly used generally for zinc. It is now not current.

Spent liquor   *See* Spent pickle liquor.

Spent pickle liquor (spent liquor)   Waste acid after its use for pickling (qv). It used to be dumped anywhere but today is chemically neutralized before disposal.

Spheroidal graphite iron (SG iron) (nodular iron)   A cast iron in which the graphite (carbon) is present in the form of nodules or spheres. It has greatly increased strength and ductility compared with ordinary cast iron.

Spiegeleisen   *See* Ferro manganese.

Spilly places (obs)   Spongy or irregularly spotted parts in wrought iron, especially sheet, caused by imperfect puddling, part of the iron having been oxidized when coming to nature (qv).

Spinner   A cavity in a mould runner system designed to make the molten metal swirl and throw out slag and dirt before the metal enters the mould cavity proper.

Spirit plate (obs)   *See* Wind wall.

Splash   *See* Shell.

Splash can   A thin sheet steel open-topped cylinder placed in the bottom of a top-poured steel ingot to contain the first stream of metal and prevent it from splashing on the walls of the mould. It melts into the ingot as the mould fills up.

Splaying   A separate cold-rolling operation given to hoops after hot rolling to provide the necessary camber to fit barrels etc.

Split pattern   A foundry pattern made in two or more parts.

Split-wind blowing   Supply of blast to two or more blast furnaces from a common main, or an extra supply of blast to a furnace already blown from another source.

Sponge iron   An iron produced by direct reduction from high purity iron ore, especially in Sweden. It is usually in lumps or cakes of porous form and is used as the raw material for some qualities of steel.

Sponging slag   *See* Foaming slag.

Spool roller A roller in a roller table which has flanges to guide the piece.

Spoon (sampling spoon) (say ladle) A steel ladle a few inches diameter fixed to a long steel handle. It is used for taking samples of molten steel from a steel furnace for analysis. The spoon is dried in the furnace, coated with slag and dipped into the molten metal to take the sample.

Spoon sample *See* Bath sample.

Spoon sleeker A moulding tool with a spoon-shaped end, used for smoothing radiused parts of a mould.

Spout (launder) A refractory-lined trough for conveying molten metal, eg for pouring molten iron into an open-hearth furnace. Some troughs which are truly spouts have names of their own, eg on a blast furnace, where the tapping spout is called the runner. A portable spout, lifted into position by the shop crane, is used for charging molten iron into an open-hearth furnace. A fixed spout is sometimes called a launder. The tapping spout of an open-hearth furnace is usually called a launder.

Spray cooling Cooling of the hearth and well of a blast furnace by water sprays directed at the outside of the jacket. Also used for the lower part of cupolas.

Spray refining Another name for spray steelmaking. *See* Continuous steelmaking.

Spray steelmaking *See* Continuous steelmaking.

Spray tuyere A hollow tuyere into which the cooling water is introduced in the form of a spray.

Spread The amount by which a sheet, plate, or strip increases in width (as distinct from length which is the major increase) during rolling.

Sprig rods (obs) *See* Slit rods.

Sprigs Small nails used to reinforce the sand in a foundry mould.

Spring hammer A power hammer driven by a crank or eccentric, with a coil or half-elliptic spring interposed between the drive and the tup. It enabled a firm blow to be given to work of varying thicknesses, the spring accommodating the difference.

Spring necking tools *See* Necking tools.

Spring oliver *See* Oliver.

Spring, roll

Spring, roll   *See* Roll spring.
Spring steels   Steels made, as the name indicates, for making
springs. They range from plain carbon steel to silico-manganese
and chromium-vanadium steels.
Spring swage   *See* Swage (1).
Sprue (downgate)   In a foundry mould the channel which leads
from the gate to the mould cavity. The term is also used to
describe the metal which solidifies in the sprues and has to be
cut off the casting, and to describe the downgate itself.
Sprue cutter   A metal tool used to cut the sprues in foundry
moulds.
Spun cast   *See* Centrifugal casting.
Squabbing rolls (stepped rolls)   A pair of rolling mill rolls in
which the passes are rectangular, formed by turning a series of
steps in the roll barrels.
Square pass (diagonal pass)   A rolling mill pass for rolling
squares. It is formed in the rolls with the diagonals vertical and
horizontal, hence the alternative name diagonal pass.
Square root angle   Rolled angle sections normally have a radius
inside at the root but they can be rolled with the root square.
They are more expensive and not used generally – only when
the end use makes it unavoidable.
Squares   Steel bars hot rolled or cold drawn to a square cross
section.
Squeezer (obs)   A machine for expelling the slag from puddled
balls by squeezing instead of hammering. There were several
types, of which the crocodile or alligator squeezer was the
commonest. Winslow's used a revolving cam.
Stabilized wire   Steel wire treated by a British proprietary
process involving heat treatment under tension. It produces wire
with physical properties superior to that which is normally
stress-relieved.
Stack (1) (shaft)   That part of a blast furnace from the top of
the boshes to the throat or charging point.
Stack (2) (shaft)   That part of a cupola from the tuyeres to the
charging door.
Stack batter   The taper or angle by which the stack of a blast
furnace widens out towards the bosh. It has varied greatly over
the years and still does, though about 1 in to 1 ft could be said

to be typical today.

Stack cooler   A hollow, flat copper casting, numbers of which are built into the stack lining of a blast furnace and connected to a water cooling system.

Stack moulding   *See* Multiple moulds.

Stack of moulds   *See* Multiple moulds.

Staff (1)   *See* Porter bar.

Staff (2) (obs)   An iron bar used to manipulate a puddled ball under a helve hammer. It was heated to a welding heat and welded to the ball with a blow of the helve. When the ball was shingled to a bloom the staff was cut off by the shingler.

Staff carrier (obs)   A boy who heated the staffs and took them to the shingler.

Staff, hearth   *See* Hearth staff.

Staff hole (obs)   A small hole in the front plate of a puddling furnace, between the firebridge and the firedoor, used for heating the ends of staffs. Often found on puddling furnaces long after the steam hammer had made the staff obsolete.

Staffordshire iron (obs)   Wrought iron made in Staffordshire from pig iron made from local ores.

Staffordshire marked bars (obs)   *See* Brand.

Staffordshire oven (obs)   An old form of hot blast stove (qv).

Staffordshire tuyere (obs)   *See* Tuyere.

Stage   The shop floor on the charging side of a steel-melting furnace.

Stainless iron   *See* Stainless steels.

Stainless steels (rust-resisting steels)   Stainless steels are of three main types.

(1) Ferritic; less than 0.10% carbon, and from 11 to 30% chromium. This group cannot be hardened by heat treatment. Those containing from 12 to 14% chromium are usually called stainless irons.

(2) Martensitic; more than 0.10% carbon and 11 to 20% chromium, nickel up to 3%. Can be hardened and tempered.

(3) Austenitic; minimum chromium and nickel, 11 and 8% respectively, total nickel and chromium must be at least 23%. *See also* Heat-resisting steels.

Stall (obs)   A brick-built wall forming a rectangle, open at the

Stamp (obs)

front, used for calcining iron ore. Stalls were built in groups.

Stamp (obs)   A piece of iron from the refinery.

Stamping   *See* Drop forging.

Stand (1) (obs)   A thick clay disc placed between the crucible and the firebars of the furnace. It kept the crucible well up above the comparatively cold firebar area and right into the fire.

Stand (2)   A pair of rolling mill housings complete with rolls, housing screws, chocks, driving pinions, etc. Two or more stands could be connected to form a train.

Standard   *See* Housing.

Standard area of tinplate (SAT)   The current basis for tinplate pricing. It replaces the old basis box (qv). SAT is 100,000 in² and the boxes are made up in units of 100 sheets or multiples of 100. 'Substance' (qv) is no longer used, thicknesses being stated in decimals of an inch.

Standard square   A refractory brick 9 × 4½ × 3 or 2½ in.

Standing plate (obs)   The cast-iron floor plate in front of a puddling furnace where the puddler stood to work.

Stannington clay   A natural fireclay found in the neighbourhood of Sheffield and formerly used as a constituent of crucibles.

Star, rumbling   *See* Rumbling star.

Stars (obs)   *See* Fence iron.

Stassano furnace (obs)   An arc furnace of Italian origin, designed for smelting iron and also for steel making.

Stave coolers   Large, heavy, hollow iron castings, incorporating steel water pipes, built into the outside of the hearth jacket of a blast furnace to cool the refractories.

Staybolt iron (obs)   Wrought iron bars of high quality specially made for stays in locomotive boilers.

Staybrite steels   A British proprietary range of stainless steels.

Steam drop stamp (obs)   A drop hammer which was lifted by a steam mechanism, eg Brett's patent lifter (qv), and fell by gravity. Not to be confused with a steam stamp, which had power on the falling stroke.

Steam ejector   A steam operated device for producing a vacuum in large vacuum-degassing vessels. Several ejectors are usually employed in series. They exhaust into a condenser. Cf. booster ejector.

Steam hammer (Nasmyth hammer)   A mechanical hammer with a piston raised and driven down by steam. Used for shingling and forging but now largely superseded by hydraulic and other presses. Often called a Nasmyth hammer, after the inventor, but not all steam hammers are of the Nasmyth pattern so the usage can be misleading.

Steam-hydraulic press   A hydraulic press in which the forging pressure is produced by a large-diameter steam cylinder coupled direct to a small-diameter water hydraulic cylinder, both being mounted vertically on top of the press. The actual steam/hydraulic cylinder assembly is called an intensifier.

Steam-oxygen process   This is really the VLN process (qv).

Steam stamp (obs)   A drop stamping hammer in which the tup lifting is done by a direct-acting steam cylinder mounted over the tup. It was really only a specially adapted steam hammer. Not to be confused with a steam drop stamp (qv).

Steam tempered   Steel treated in a furnace atmosphere of steam, which produces a very thin (about 0.0002 in thick) surface layer of blue-black oxide. This layer retains lubricant and the process is sometimes used on engineers' cutting tools for this reason.

Steckel hot mill   *See* Steckel mill.

Steckel mill   Strictly a four-high mill (qv) for cold rolling wide or narrow strip, in which the rolls are not power driven, the metal being pulled through by powered coilers. But the term is also applied to a form of four-high hot mill in which both the rolls and the coilers are driven. In a few Steckel hot mills the coilers are in box furnaces, called hot boxes.

Steel   Steel is defined in more than one way. A general definition is that it is a malleable alloy of iron and carbon, the carbon not exceeding about 1.7%. If other elements are present in appreciable quantities the steels are alloy steels (qv).

Steel body   *See* Body.

Steel, cast   *See* Cast steel (1) and (2).

Steel colour   *See* Temper colour.

Steel conversion factors   In the manufacture of any steel product there are losses due to the necessary cropping, dressing, and trimming; a ton of steel in the furnace will not make a ton of finished product. There are accepted factors which give the

Steel facing

amount of finished product obtainable from a unit quantity of steel and, conversely, the amount of steel needed to make a unit quantity of finished product. They vary according to the nature of the product but the average of all products is obtained as follows:

for ingot to product multiply by 0.730;

for product to ingot multiply by 1.370.

Steel facing   *See* Steeling (1).

Steeling (1) (obs) (steel facing)   Welding a hardenable steel face to a piece of wrought iron (obs) or mild steel for making edge tools. Virtually obs.

Steeling (2) (obs)   A former alternative name for carburizing (qv).

Steeling (3) (obs)   Charging the raw materials into a crucible or pot by pouring them down a metal funnel.

Steel melter   *See* Melter.

Steel mill   *See* Mill.

Steel, semi   *See* Semi-steel.

Stelco process (1)   A Canadian proprietary process for controlled cooling of hot rolled, coiled steel bar and rod.

Stelco process (2)   A direct-reduction process using iron ore pellets, natural gas, and coal in a rotating kiln. The product is sponge iron.

Stelvetite   *See* Plastic coating.

Stepped rolls   *See* Squabbing rolls.

Sticker   An ingot which is difficult to remove from the mould.

Stickers (obs)   Hot, pack-rolled sheets which have stuck together during rolling and have to be opened (qv).

Stiffeners   Extra bearings applied to the top roll of a section mill when the roll shape has deep grooves in it. They are actually half-bearings, carried by a beam across the top of the mill housings.

Stirling's toughened iron (obs)   A mixture of cast and wrought iron, melted together and supposed to make a tougher product than ordinary cast iron. Not much used.

Stitching   An alternative to welding for joining strip passing through continuous processing such as annealing. A mechanical press punches the surface of one strip into the other. The stitching is cut out after processing.

Stock converter   A form of Bessemer converter in which the charge (usually scrap) is melted down by an oil burner before blowing. Now rare.

Stock hearth (obs)   A hearth formed at floor level against a small refractory brick wall and used to heat large pieces of wrought iron for forging. The fuel was small coal. A blow pipe passed just through the wall at floor level and the tuyere was formed by pushing an iron bar into the pipe, building up wet small coal around the bar which was then withdrawn and the fire was lighted at the end of the tuyere so formed. The tuyere burnt back gradually and was remade each day before starting work.

Stock line   The point in the stack of a blast furnace to which the charge reaches.

Stock rod (test rod)   One of several rods descending vertically in the top of a blast furnace to rest on the charge and so indicate its height.

Stock-rod cover (obs)   *See* Poker-hole cover.

Stocktaker   A workman in charge of the blast furnace pig beds and responsible for disposing of the iron when cast.

Stone   *See* Iron ore.

Stopper (1)   A refractory-covered steel rod used in bottom-pour ladles to start, stop and control the flow of steel from the ladle. It works into a refractory nozzle set into the bottom of the ladle. Stoppers are also used in tundishes.

Stopper (2)   *See* Peg.

Stopper-drying oven   A gas-fired oven for drying steel ladle stoppers (qv) before they are fitted to the ladle.

Stoppered tundish   A tundish used between the ladle and the mould in continuous casting. It has a stopper by which the steel flow into the mould is controlled either manually or mechanically from a radio-active sensing device which monitors the metal height in the mould.

Stopper gear   The handle and levers used to control a ladle or tundish stopper.

Stopper hole (obs)   A small hole at the bottom of the puddling furnace working door, used for inserting the tools without opening the door.

Stopping (1) (v) (tamping)   Closing the taphole of a furnace

Stopping (2) (n)
with clay or other refractory.

Stopping (2) (n)   The refractory plug in the taphole of a blast
furnace. It is removed to to tap the furnace.

Stopping off   A method of moulding to produce a casting
shorter than the pattern. The mould is made in the normal
way and the part or parts of the mould cavity not required are
filled with moulding sand. Stopping off is usually applied only
to simple shapes such as pipes.

Stops   *See* Billet lengths.

Stourbridge clay   A natural clay found near Stourbridge,
Worcestershire, and formerly used extensively for making
refractory bricks and as a constituent of crucibles.

Stove-cleaning gun (obs)   A special form of gun like a small
cannon on wheels, with the mouth pointing upwards. It was
used when crude gas was burnt in hot-blast stoves. The mouth
of the gun was inserted into one of the stoves and a charge was
fired; this loosened the dust deposited from the dirty gas.

Stove minder (hot-blast man)   A man in charge of the hot-blast
stoves at a blast furnace; responsible for keeping the blast at the
required temperature.

Straight   Sometimes used in the sense of 'ordinary' as in straight
carbon steel as distinct from alloy steel.

Straight-line firing   A method of firing a soaking pit (qv).
Burning fuel enters through a port in one end wall, passes
straight through the chamber and exhausts through a port in the
opposite end wall. Cf. loop firing, umbrella firing.

Strand (1)   The travelling hearth of a sinter machine.

Strand (2)   The travelling line of moulds on a pig-casting
machine.

Strand (3)   The pass line in a continuous bar or rod mill; many
have two lines of passes side by side in each stand of rolls and
are then called a two-strand mill. Some may have more than
two strands.

Strand (4)   The mould, cooling, withdrawal, and cut-off
apparatus of a continuous casting machine (qv). Such machines
can be single-strand or two-, three-, or more strand. The word
also applies to the actual piece of metal being cast, which does
not become a billet, bloom or slab until it is cut off.

Strand annealing   Another name for continuous annealing (qv).

Strand pass   The last pass but two in a hand rolling mill. *See also* Leader pass.

Strategic Udy process   A direct-reduction process. Iron ore, flux and carbonaceous matter are fed to a rotating kiln and the pre-reduced product is discharged to an electric furnace for final reduction.

Strawberry blister   *See* Pickling patch.

Stream   A common term from the petro-chemical industry is finding its way into iron and steel making; a blast furnace or even a complete steelworks is said to be 'on stream' when it is fully operational. Its use in this context is totally unnecessary and should be discouraged. The iron and steel industry is adequately provided with terms of its own.

Stream degassing   Another name for ladle-to-ladle degassing (qv).

Stress relieving   A heat-treatment process, in which the object is raised to a suitable temperature and held there for a specified time to allow internal stresses, caused by work done on the object, to 'creep' away.

Stretch-forging machine (swing hammer machine)   A machine having eight forging hammers arranged in four pairs in a heavy housing; each pair works in a different plane. The hammers swing inwards to strike the workpiece and reduce its cross-sectional area by a combination of forging and stretching.

Strickle (loam board)   *See* Loam moulding.

Strickle work (strickling)   Using a strickle to form foundry moulds without a pattern. Generally done in loam (qv) but some sand moulds (eg for flywheels) can be wholly or partly made by strickling.

Striker   The blacksmith's assistant who strikes with a sledge hammer as directed and does other jobs as required.

Strip (1)   Iron or steel rolled out into long, thin, flat strips. Iron is no longer rolled. Steel up to about 24 in wide is strip or narrow strip, above this, wide strip (qv). The dividing line is sometimes said to be 18 in, but 24 in is more generally accepted.

Strip (2) (v)   To remove an ingot from its mould.

Stripper   *See* Ingot stripper.

Stripper crane   *See* Ingot stripper.

Stripping bay   A part of a main steelworks building (or a separate building) devoted to ingot stripping.

Stripping plate   *See* Guide.

Structural mill   A rolling mill for producing rolled structural sections.

Structurals   *See* Structural sections.

Structural sections (structurals)   Hot rolled steel sections in the form of angles, beams, channels, tees, joists, etc.

Stub   The end of an electrode left after use in an electric-arc or consumable-electrode-refining furnace.

Stückofen (obs)   A primitive form of furnace formerly used in Germany. It was really neither a blast furnace nor a bloomery but had some characteristics of both.

Substance (obs)   The thickness of tinplate expressed as weight per unit area. It was actually expressed in pounds per $31,360$ in² of tinplate, ie pounds per basis box (qv). Common substance was 108 lb per basis box and was called IC.
See Standard area of tinplate.

Suction box   *See* Wind box (3).

Sulphur   A common element, symbol S, which is generally deleterious in steel, though it may be added deliberately to free-cutting steels (qv).

Sulphuric acid   A strong acid, formula H_2SO_4, used diluted for pickling (qv) steel.

Sulphur print   A means of examination for sulphides in steel. The surface is polished and a bromide paper, soaked in dilute sulphuric acid, is placed in contact with it. A dark brown stain on the paper shows the position of the sulphides, and also indicates the flow lines in a forging.

Superheat   To heat metal beyond the temperature required in order to allow for some temporary condition, eg when molten steel is to be vacuum degassed.

Surface hardening   *See* Case hardening.

Super beneficiation   *See* Beneficiation.

Supergrip plates   *See* Anti-slip plates.

Super Tread plates   *See* Anti-slip plates.

Suspended core   *See* Hanging core.

Swage (1) (n)   A forging tool with a convex surface, used for finishing round surfaces. For power hammer work swages are often fixed together in pairs by a long hairpin-type spring; these are spring swages.

Swage (2) (v)   To form steel by a succession of blows by dies revolving around it.

Swage block   A block of cast iron with perforations of various shapes and sizes right through it and grooves of various shapes around its edges. It complements the blacksmith's anvil. The grooves can be used for shaping and the holes can support various tools.

Swarf   Small particles of metal such as cast iron, produced by machining and remelted, usually in electric furnaces, for re-use.

Swarf, wheel   *See* Wheel swarf.

Swealing   *See* Wash heating.

Swedish Lancashire iron   Wrought iron made in Sweden from Swedish pig by the dry-puddling process.

Sweeling   *See* Wash heating.

Sweep (sweepy)   A rolled piece, faulty because it has a sweeping curve on one side, usually caused by uneven heating.

Sweep moulding   *See* Loam moulding.

Swell   A defect in a casting caused by the pressure of the molten metal displacing the sand of the mould.

SWG   *See* British Standard Wire Gauge.

Swift   A support for a coil of rod or wire, used in wire drawing.

Swing grinder   A heavy grinding machine suspended from above and guided by handles over the surface to be ground. Used for fettling castings and for dressing billets, blooms, and slabs.

Swing-hammer machine   *See* Stretch-forging machine.

Swording (obs)   *See* Opening.

Syphon pipe (obs)   The air pipe in a Staffordshire hot-blast oven or stove.

T

Taggers (obs)   Very thin tinplates often used for making tags.

Tagging   Pointing the end of a bar or wire to enable it to enter the die for drawing.

Tail scale   *See* Trickle scale.

Talbot process   A steel-making process using pig iron in a tilting open-hearth furnace. Only about one-third to half the molten metal is tapped, leaving sufficient to dilute the impurities in the next charge of pig iron.

Tamping   *See* Stopping (1).

Tandem mill   *See* Continuous mill.

Tandem rolling   Feeding one ingot, bloom, or billet into a rolling mill so that its entry end touches the back end of the piece already in the mill.

Tap (v)   To let iron, steel, or slag flow from a furnace by removing the stopping which holds it back or, in some modern furnaces, by tilting the furnace mechanically.

Tap cinder (obs)   Slag tapped off a puddling furnace.

Tap degassing   A process for degassing molten steel. The steel is tapped direct from the furnace through a tundish or pony ladle into an evacuated ladle, which is then teemed in the usual way.

Taphole (iron notch – rare)   The hole at the front of any furnace melting iron or steel, through which the molten metal is allowed to flow as required. On a fixed furnace the taphole is normally stopped with refractory material until tapping is required, but on a tilting furnace it may be open, being above metal level until the furnace is tilted. On a blast furnace the taphole is sometimes called the iron notch. Large cupolas often have open tapholes and tap continuously.

Taphole drill   A mechanical (electric or compressed air) drill used to start the tapping of a blast furnace. On large furnaces it is remotely controlled.

Taphole gun   *See* Mud gun.

Tapping bar   An iron or steel bar with a pointed end, used for tapping.

Tap steel   A high-carbon or high-carbon low-alloy steel used for making engineers' thread-cutting taps.

Tap-to-tap time   The steel making cycle, ie the time from tapping a heat of steel to the time when the next charge is tapped.

Tar   *See* Coal-tar fuels.

TC   *See* Graphitic carbon.

Teapot ladle   A ladle used in iron foundries. It has a spout not unlike that of a teapot but parallel with the side of the ladle proper. Molten metal flows from the bottom, up and out of the spout as the ladle is tilted; slag floating on the metal does not enter the spout.

Tear (pull)   A fault in a casting caused by it having been restrained while cooling so that the metal could not contract normally and tore as it solidified.

Teazer   *See* Slagger.

Tee   A structural rolled section shaped in cross section like a letter T.

Teem   To cast molten steel from a ladle into an ingot mould. Teeming is basically of three types: (1) Direct teeming from the ladle into the open top of the ingot mould; (2) Tundish teeming, when a tundish is placed between the ladle and the mould to give a constant head of metal; and (3) Indirect or uphill teeming, when the metal passes first to a trumpet or runner leading to the bottoms of a group or nest of ingot moulds, in which the metal rises. Method (3) is also called bottom casting or bottom teeming. Most teeming is done from a bottom-pour ladle but the term was also applied to pouring from a crucible direct into the mould in crucible steel making.

Teemer (1)   The man who does the ingot teeming.

Teemer (2) (obs)   A key man in the crucible steel making team. He poured or teemed the molten steel from the crucible into the ingot mould and was in overall charge of the steel making process.

Teeming hole (obs)   A long narrow slot in the floor in which the ingot moulds stood for teeming from crucibles.

Teeming line   The height to which molten steel is teemed into ingot moulds.

Temper (1) (v) (let down)   To reduce the degree of hardness of steel to the point required for service.

Temper (2) (n)   The state of hardness in steel after tempering.

Temper (3) (v) (obs)   To prepare clay for making crucibles by treading it with the bare feet.

Temper (4) (obs)   Crucible steel was divided into grades known as tempers (actually the carbon content), by visual examination of the fracture. The practice varied but a typical grading was from No 1 temper (0.5% C) to No 6 temper (1.5% C). A seventh temper (about 2% C) was offered by some makers. The temper names were: 1 spring heat, 2 country heat, 3 single shear heat, 4 double shear heat, 5 steel through heat, 6 melting heat, 7 doubly converted. These are typical names only — practice varied.

Temper (5) (obs) (season)   To allow a clay crucible to dry slowly and naturally before firing or annealing it.

Temper (6) (v)   To condition foundry moulding sand after use or to treat new sand to fit it for use, by adding other materials such as coal dust, and mixing it mechanically, is often called tempering it.

Temperature measuring colours   A method of approximate measurement of surface temperature. Metal salts mixed with alcohol into a paste and made into coloured crayons with wax, are used to make a mark on the surface of a piece of metal. When the metal reaches the temperature represented by the particular crayon, the colour of the mark changes. Temperatures from about 45 to 926°C are covered by a standard range. Tempilstik is a proprietary temperature-indicating crayon.

Temper brittleness (Krupp's disease) (obs)   Brittleness found in some medium- or low-alloy steels under certain conditions of heat treatment. It was formerly called Krupp's disease after the German firm which first detected it.

Temper colour   The colour of the oxide layer which forms on
bright steel when it is heated. The colour ranges from light
straw at about 210°C to grey at about 330°C. Colour used to
be used widely to determine tempering temperatures but
instruments are used today. Coloured charts were issued by tool
steel makers showing the temper colours — they can be found
today.
Cf. heat colour.

Temper rolling   *See* Skin pass.

Tempilstik   *See* Temperature-measuring colours.

Tension bridle   *See* Drive bridle.

Tension reel   A strip coiler which applies tension to the strip as
it leaves the rolling mill and coils it up.

Terne plate   Thin sheet steel, hot-dip coated with a lead-tin
alloy, the tin content being up to about 25%.

Test bar   *See* Test piece.

Test lug   *See* Test piece.

Test piece (test lug) (test bar)   A lug cast on a steel casting and
cut off for testing when the casting is fettled.

Test rod   *See* Stock rod.

Texturized steel   *See* Rigidized steel.

Theoretical mill output   The maximum theoretical output of a
rolling mill can be calculated by taking the maximum practicable
speed at the finishing stand, multiplying it by the section area
and converting the figure to tons. In practice the output will be
less than the theoretical because the mill will not be available
100% of the time for various reasons. The nearer the actual is to
the theoretical, the higher the overall efficiency of the mill and
its auxiliaries, some of which may be responsible for the mill
not working at times.

Thermal shock   Rapid heating and cooling. Some steels are
made to withstand this condition in service.

Thermostat   A device for maintaining a furnace at a pre-set
temperature by controlling the heat input automatically.

Thimble section   *See* Specials.

Third-hand melter   *See* First-hand melter.

Thirty carbon   *See* Fifteen carbon.

Thomas iron   The name given on the Continent to phosphoric
pig iron suitable for use in the Thomas process.

Thomas process   The Continental name for the basic Bessemer process (qv). Sometimes used in the UK as well.

Thomas steel   The Continental name for steel made by the basic Bessemer process.

Three-high mill   A rolling mill in which there are three rolls in each stand, mounted vertically, one above the other. *See also* Four-high, two-high.

Three-part system   *See* Hot-pack rolling.

Throat (mouth)   The narrowest part of a blast furnace, at the top of the stack.

Throat armour   *See* Armouring.

Tigershoild method   A hot-topping method (qv) of Swedish origin in which a sand head is formed in the head of an ingot mould each time it is used.

'Tight' (obs)   *See* 'Bare'.

Tight coat   A galvanized or tinned coating which is as free as possible from flaws and completely covers the base metal.

Tilted iron (obs)   Wrought-iron bars welded and forged into a longer piece under the tilt hammer. The term is interchangeable with faggoted iron (qv) except that faggoted iron could be made under any hammer and not just a tilt.

Tilted steel (obs)   Cemented steel forged under the tilt hammer. The term survived after the tilt hammer had been replaced by the steam hammer.

Tilter   Originally the operator of a tilt hammer, but the term has survived in the Sheffield area to mean a man or a firm which does hammer cogging or forging down by power hammer.

Tilt hammer (obs)   A mechanical hammer for delivering light rapid blows. Always of the first-order lever form with the fulcrum in the middle, hammer at one end and cams striking at the other. Usually had a spring beam over the hammer head to assist the blows or a recoil block under the tail, in which case the elasticity of the haft gave the spring action. The haft is correctly known as the helve, but this causes confusion with the helve hammer proper (qv). Tilt also called loosely a trip hammer and a water hammer.

Tilting   Originally the work of hammering under a tilt hammer but the term survived in the Sheffield area to mean the cogging or drawing down of steel bars under any kind of power hammer.

Tilting fingers

Tilting fingers   Mechanically operated arms, working through a mill roller table, to tilt or turn up the piece as required.

Tilting furnace   Any melting furnace which is arranged to be tilted mechanically or hydraulically for tapping. Some open-hearth furnaces are of the tilting type and all electric arc furnaces tilt.

Tilting table (lifting table)   A mechanically operated table fixed at one or both sides of a three-high rolling mill to lift the piece up from the middle-bottom to the top-middle pass, or to lower it in the opposite direction. The table is pivoted at the point furthest from the mill stand and tilts about this point.

Timp (obs)   *See* Tymp.

Tin   A metallic element, symbol Sn, which has a good resistance to corrosion and is non-poisonous. It is used extensively for coating thin steel sheet (tinplate).

Tin bar (tinplate bar)   Flat rolled steel bar specially rolled for cutting into lengths and rolling to thin sheets for making tinplate.

Tin house (obs)   The building in which tinning of tinplate was done.

Tinman (obs)   The man who worked at the tin pot and did the actual tinning by dipping the sheets to make tinplate.

Tin mill (obs)   An alternative name for a tinplate rolling mill.

Tinned iron (obs)   *See* Tinplate.

Tinplate (white iron)   Thin steel sheet with a very thin coating of metallic tin. Tinplate was formerly iron sheet. The old term white iron has long been obsolete.
See also Basis box and Standard area of tinplate.

Tinplate bar   *See* Tin bar.

Tinplate sizes (obs)   *See* Basis box.

Tin pot (obs)   A vessel containing molten tin into which sheets were dipped for tinning. The later pots could be single sweep (with only one pot of molten tin) or double sweep (with two separate pots working in tandem).

Tip iron (obs)   *See* Specials.

T-iron (obs)   Wrought iron rolled in the form of a letter T.

Todgers (obs)   Used in laying up (qv).

Tolerance   *See* Rolling margin.

Tommy   An oliver (qv) with a pair of swage tools permanently

fixed in the hammer and anvil. The tommy was really a tool of the chainmaker, but could be found in works smithies.
Cf. dolly.

Tommy hop (obs)   A water balance lift for elevating charges to the top of a blast furnace.
Cf. wind hop.

Tong hold (bar hold)   The end of a bar or forging which has been forged down to enable it to be gripped by tongs during subsequent forging.

Tonnage oxygen   High-purity oxygen produced in large quantities particularly for oxygen steelmaking, but also for other steelworks purposes such as cutting or deseaming. A ton of oxygen = 26,500 ft³

Tonnage steels   A general term for steels made in large quantities, as distinct from special and high-alloy steels.

Tool steels   Steels made specially for engineers' cutting tools. There are many specifications, ranging from plain carbon to alloy, according to end use.

Top degassing   Another name for the Ruhrstahl-Heraeus process (qv).

Top gas   *See* Blast furnace gas.

Top hat furnace   *See* Bell furnace.

Top-middle   In a three-high mill a pass formed in the top and middle rolls. A pass in the middle and bottom rolls is a middle bottom pass.

Topping (obs)   Knocking the end off a crucible steel ingot with a heavy sledge hammer, so that the teemer can judge the quality by the appearance of the fracture, and so decide the temper (qv).

Top tackle   Equipment on a rolling mill which assists the top guards to bear down on the top roll and so prevent collaring (qv). It usually includes springs or weights.

Top teeming   *See* Teem.

Torpedo ladle (cigar ladle) (Pugh ladle)   A large cylindrical ladle not unlike a torpedo in shape, mounted on a rail carriage and used to carry molten iron from the blast furnaces to the steelworks. Also called a cigar ladle, and Pugh ladle.

Torr   A measure of the vacuum used in vacuum processing. A torr = 1 mm of mercury, ie the pressure is sufficient to support

Total carbon

a column of mercury 1 mm high. For very low pressures the millitorr (0.001 mm) or micron is used

Total carbon *See* Graphitic carbon.

Total reduction *See* Reduction (2).

Toughness The property, combining strength and ductility, which enables steel to resist shock, bending, or twisting.

Track time The time between teeming of an ingot and charging it into the soaking pit.

Train (mill train) (train of rolls) Two or more stands of rolls coupled together.

Train of rolls *See* Train.

Tram lines Overfill (qv) appearing as two parallel lines on rolled bar.

Tramp element (incidental element) An element present in a steel or alloy which has got there by accident.

Tram rails Light section rolled steel rails of the railway type, made for industrial tramways. Should not be confused with tramway rails (now obs in Britain) rolled for street tramways.

Tramway rails (obs in Britain) Steel rails specially rolled for street tramways. Not to be confused with tram rails (qv).

Transfer bank (transfer bed) A cooling bank or bed with mechanical transfer devices to move the pieces sideways as they cool and eventually to push them off the bank for disposal.

Transfer bed *See* Transfer bank.

Transformer iron A low carbon, high silicon, sheet steel (not iron) made for electrical purposes. Although the material is steel the term 'iron' is often used.

Transformer sheets Steel sheets rolled specially for electrical purposes. They have a very low carbon content but high silicon (about 4%), which gives the required electrical properties.

Tray and waiter iron (obs) Thin sheet iron rolled for the makers of trays and 'waiters'.

Tread *See* Treader.

Treader (obs) A clay worker who formerly prepared the clay for crucibles by spreading it in a wooden frame on the floor and treading it for several hours with his bare feet.

Treading *See* Treader.

Trebles *See* Sheet (2) and Sheet sizes.

Trickle scale (obs) (tail scale) A sheet fault, caused by scale flaking from the ends of a pair or pack of sheets and trickling in between the surfaces during rolling.

Trim   To remove the flash from a forging by forcing the forging through suitably shaped dies.

Trimming press   A simple mechanical press for trimming the flash off drop stampings. The stamping is pushed through a die by the press punch, leaving the flash behind.

Trip hammer (obs)   A name loosely applied to both the tilt and the helve hammer (qv); in the latter case incorrectly.

Triples (obs)   *See* Sheet (2).

Triplex process (rare) (obs)   Making steel in three furnaces in stages. Pig iron was partly refined in an acid Bessemer converter, transferred to an open-hearth furnace for refining, and finally transferred to an electric-arc furnace where final refining and alloying were done.

Trompe (obs)   *See* Catalan forge.

Tropenas converter (side-blown converter)   A type of Bessemer converter in which the tuyeres are on the side of the vessel instead of in the bottom. Now rare.

Troughing (bridge deck)   Hot rolled sections in the form of a trough, often used as bridge decks, hence the alternative name.

Trumpet   *See* Teem.

Trunnions   Pivots on which a ladle or converter turns.

Tu-arn   *See* Tuyere.

Tub (obs) (blowing tub)   A blowing cylinder with a reciprocating piston, operated by a steam engine or a water wheel.

Tu-iron   *See* Tuyere.

Tundish teeming   *See* Teem.

Tungsten (wolfram)   A metallic element, symbol W, used with other elements in alloy steels. Also called wolfram, though strictly that is the ore of tungsten.

Tungsten steels   Alloy steels containing, typically, 12 to 20% tungsten, 1 to 4% vanadium, 0.7 to 1.2% carbon. Very hard and tough.

Tunnel (obs)   An old name for the throat (qv) of a blast furnace — hence tunnel head (qv).

Tunnel furnace   A long annealing furnace of tunnel shape, through which the material to be annealed is transported on

Tunnel head (obs)

mechanically propelled cars.

Tunnel head (obs)   A short chimney stack (usually brick) at the top of a hand-charged blast furnace to carry the fumes above the filling place. It had from one to six or more apertures or filling holes to permit charging.

Tunnel heads were common but not universal; many furnaces had only a low rim round the top against which the barrow wheels rested as the load was tipped.

Tup (1)   The actual hammer part of a steam hammer; it has a detachable and renewable face or pallet. The weight of the tup is the nominal size of the hammer.

Tup (2)   The falling part of a drop hammer.

Tup pallet   *See* Pallet.

Turbo blower   A blast furnace blower in which the air is compressed in stages in a turbine. It can be driven by a steam turbine or an electric motor. Gas turbine drives have been tried.

Turk's head   An arrangement of four undriven rolls mounted in a housing in the form of a square, so that they bear simultaneously on all four sides of a piece of metal being drawn through. Used for shaping wire or cold-drawn steel bars.

Turn (obs) (n)   A complete working shift or turn of duty. When the two-shift system (12 hours per shift) was in operation the turns were 6 am to 6 pm (day turn) and 6 pm to 6 am (night turn). With the coming of the three-shift system (8 hours per shift) the word turn tended to pass out of use and the shifts are so called. They are; 6 am to 2 pm (morning shift), 2 pm to 10 pm (afternoon shift) and 10 pm to 6 am (night shift). There are local name variations.

Turning gear   *See* Burden chain.

Turn-over machine   *See* Roll-over machine.

Turn pegs   Mechanically operated fingers working in the entry table of a plate rolling mill to turn the plate after broadsiding (qv).

Tuyere (tu-iron) (tu-arn) (tweer) (twire)   Also spelt tu-iron, tu-arn, tweer, twire, etc. The end of a blast pipe conveying blast to a furnace or hearth. Always detachable for easy replacement. Before hot blast, tuyeres were not usually cooled but the increased working temperature of hot blast called for water cooling; hence water tuyere (obs). The Staffordshire tuyere (obs) consisted of two short truncated cones one within the other with a space between to circulate cooling water. Condie's or the Scotch tuyere (obs) was of cast iron with a conical wrought iron pipe cast in for the water. The tuyere over the taphole of a blast furnace is the monkey tuyere. Modern blast-furnace tuyeres are hollow copper castings.

Tuyere arch (obs)   An archway in a stone or brick blast furnace to receive and give access to the tuyere.

Tuyere cap   A small cover over the opening or wicket at the back of a tuyere stock to allow of inspection of the interior of the furnace without removing the blowpipe. The bright spot thus viewed is the 'eye of the furnace'.

Tuyere snatcher   *See* Snatcher.

Tuyere stock   A refractory-lined pipe between the blowpipe and the gooseneck on a blast furnace.

Twenty carbon   *See* Fifteen carbon.

Twenty-five carbon   *See* Fifteen carbon.

Tweer   *See* Tuyere.

Twice broken iron (obs) (rare)   A very rare alternative name for puddled iron made by the dry puddling process. It was first refined, then puddled or 'twice broken from the pig'.

Twire   *See* Tuyere.

Twisted strip   Strip which is distorted in the centre because the edges are longer than the middle. Cf. full strip.

Twist guide   A rolling mill guide (qv) which twists or turns the piece through 90° between passes, as in rolling from oval to round. Twist guides may be solid metal or formed of rollers.

Two-high mill   A rolling mill with only two rolls in each stand. *See also* Three-high, Four-high.

Tyre-fixing hammer (obs)   *See* Tyre hammer.

Tyre-fixing rolls   A special machine for the same purpose as a tyre hammer (qv) but using a pair of small rollers instead of a hammer.

Tyre hammer (obs) (tyre-fixing hammer)   A special steam or air hammer used for fixing railway wheel tyres to the wheels by closing the edge of the tyre over a retaining rim.

Tyre mill   A special-purpose machine for rolling a pierced cheese into railway tyres.

Tymp (obs) (timp) (foreplate)   The arch over the forepart of an open-forepart blast furnace.

Ultrasonics   A non-destructive method of testing iron and steel for internal soundness. Ultrasonic waves are sent into the article being tested and reflected back to a receiver. Concealed faults, such as blowholes, will reflect the ultrasonic waves and will show up on the instrument screen.

UM plate (universal mill plate)   Steel plate which has been rolled in a universal mill.

Umbrella firing   A method of firing a circular soaking pit (qv). The gas enters the chamber through a central port and the flame goes upwards and spreads out to impinge on the roof and leaves via a series of ports; the flame is therefore umbrella-shaped. Cf. loop firing, straight-line firing.

Uncoated tinplate base   *See* Full-finishing blackplate.

Uncoiler (pay-off reel)   A machine for feeding strip or wide strip continuously to further processing such as re-rolling, heat treatment, tinning, galvanizing.

Underdraught   *See* Overdraught.

Underfill   In rolling, a piece which has not filled the roll pass and is not dimensionally true. Cf. overfill.

Underhand (obs)   A puddler's assistant.

Underkilled steel   *See* Killing.

Unequal angle   *See* Angle iron.

Unequal draught   Incorrect setting of rolls, which are not parallel, causing a piece to be rolled more heavily on one side than the other.

United inches   Used in angle rolling = the sum of both legs of

Universal beam (broad flange beam)

the angle.

3 in + 3 in = 6 united inches, or 4 + 3 = 7.

Universal beam (broad flange beam)   Now the standard form of rolled steel joist or H-section. It is rolled on all faces simultaneously in a universal mill (qv).

Universal column   Similar to a universal beam but of heavier section.

Universal flats   Steel flats rolled in a universal mill, ie rolled on the edges as well as the flat faces.

Universal mill (Grey mill – obs)   A rolling mill, used for plate or sections, which has, in addition to the main rolls, another pair arranged to roll on the sides of the piece. Thus all four sides are rolled simultaneously.

Universal mill plate   *See* UM plate.

Unkilled steel   *See* Killing.

Unscrambler   *See* Descrambler.

Upcoiler   *See* Coiler.

Upcut shear   A mechanical shear in which the moving blade rises to press the piece to be cut against the top, stationary blade.

Upended forging   *See* Upset.

Upender (coil upender)   A machine for turning a coil of strip so that its axis, which is normally horizontal (in which position it is rolled and coiled), is vertical, as for charging to an annealing furnace. The same, or a similar, machine is used to turn the coil back to its original position.

Uphill teeming   *See* Teem.

Upset (v & n) (upended forging)   If in forging, a bar is to be made of greater size at the end, that part only is heated and hammered or bumped on the floor or anvil, to swell out the hot (therefore soft) part.

To swell out a part of the bar other than the end it is heated at the point required and bumped endwise on the floor or anvil. This is called jumping.

Uptake (1)   A vertical pipe leading from the top of a blast furnace to join the downcomer and lead the furnace gas away. Four uptakes are common on a modern furnace.

Uptake (2)   A vertical or near-vertical flue leading from the slag pocket of an open-hearth furnace and passing waste gases to the pocket, or heated air or gas from it to the port, according to the direction of gas flow at the time.

Use (n) (preform)   A partly-shaped piece of steel ready for finishing by drop stamping or hot pressing. Similar to American dummy.

Use-iron (obs) (n)   Unmachined or semi-finished forgings made for engineers, especially steam-engine builders.

Use rolls   *See* Gap mill.

U-tube (manometer)   A glass tube shaped like a letter U and partly filled with liquid. One end is open to atmosphere and the other end connected to an air pressure or suction pipe. The difference in levels of the liquid in the two legs of the U indicates pressure or suction. It can only be used for low pressures or suctions.

V

Vacuum-arc melting   *See* Vacuum-arc remelting.

Vacuum-arc remelting (VAR)   A consumable electrode is lowered into an evacuated, water-cooled mould and an electric arc is struck between the electrode and the mould base. As the end of the electrode melts off, gases are drawn away by the vacuum and the molten metal solidifies again in the mould. A steel refining process only – not steel making.

Vacuum casting   Casting an ingot or any other type of casting in a vacuum. Can be a form of degassing or can be a foundry process (only with high-grade alloys) where casting takes place in a vacuum chamber.

Vacuum degassing   The general name for several processes used to remove gases from molten steel.
See Vacuum casting, Ladle-to-ladle degassing, Stream degassing, DHH process, RH process, Top degassing, Continuous degassing.

Vacuum lock   A small double-gated chamber in the side of the main chamber of a vacuum melting furnace, through which alloying additions can be made, or samples of metal taken, without breaking the processing vacuum.

Vacuum melting (vacuum induction melting) (VIM)   Loosely applied to any process where metal is melted in a vacuum but should be confined to vacuum induction melting. An electric induction furnace (qv) is mounted in a vacuum chamber together with ingot or other moulds. It can be controlled, and tilted for tapping, from outside the chamber by remote control, and vacuum locks enable alloying materials to be added and samples to be taken without breaking the vacuum. Used for very high-grade alloys.

Vanadium   A metallic element, symbol V, used in special alloy steels.

VAR   *See* Vacuum-arc remelting.

Vein stuff   *See* Gangue.

Vent   A small hole in a sand mould for the escape of gas evolved when the hot metal is poured in. Several, often many, vents are used.

Vent wire   A stiff piece of wire pushed into the sand of a mould to make the vent holes.

Vermicular graphite iron   A modified form of SG iron in which the graphite is wormlike in form. It is new and still the subject of development and research.

Vermiculite   A mineral substance which exfoliates on heating. Used in foundries as a binder for sand. Also used as an insulating layer if it is required to keep a ladle of molten metal hot temporarily.

Vertical mill   A rolling mill in which the roll axes are vertical. Usually single stands are made in this way and interposed between horizontal stands in a continuous bar or rod mill.

Vessel   *See* Converter.

VIM   Vacuum induction melting (qu).

VLN process (recently obs) (oxygen-steam process) (steam-oxygen process)   A bottom-blown steel converter similar to a Bessemer converter but blown by a mixture of oxygen and steam instead of air. Used to make steel for rolling to deep-drawing sheets, where nitrogen (picked up from the air in an air-blown converter) would be deleterious.

V stockline   According to the diameter of the large bell in a blast furnace the charges tend to settle either in the form (in cross section) of a letter V or a letter M. In the V stockline the heavier particles settle around the stack walls and the lighter, larger ones, usually coke, collect in the centre. In the M stockline the coke and ores settle to the walls and the heavier ore goes to the centre.

Wabbler *See* Wobbler.

Wad A cake or disc of metal punched out of a forging during trimming. It is really part of the flash (qv).

Wagon (obs) *See* Charging wagon.

Wagon drop *See* Furnace bank.

Wagon lift *See* Furnace bank.

Wagon-tyre iron (obs) *See* Cart-tyre iron.

Waiter iron(obs) *See* Tray and waiter iron.

Walking beam A means of conveying steel bars, billets, slabs, etc across a cooling bank or through a furnace. The material to be conveyed rests on a metal grid and a second grid is arranged to lift upwards and forwards between the bars of the stationary one, so lifting the material and 'walking' it forward, before returning to make another stroke.

Walloon process (obs) A method of making wrought iron, by melting the ends off white iron pigs and allowing the molten drops to pass through an air blast, which decarburized them. The pigs were pushed forward as they melted. Charcoal was the fuel.

Wap A single turn in a coil of wire or strip.

Warted plates *See* Anti-slip plates.

Warm blast iron (obs) Pig iron made in a blast furnace with the blast at a temperature above ambient but not as hot as possible. A very imprecise term.

Warm working The forming of steel at temperatures in the

Wash heating (swealing) (sweeling)

region of 650 to 800°C. At present only used for special alloy steels (gas turbine discs, for example), this process may become of increasing importance.

Wash heating (swealing) (sweeling) Soaking an ingot at high temperature under oxidizing conditions so that the surface melts and runs off as a fluid slag. Surface defects can be removed in this way but it is expensive in material.

Wash, mould *See* Mould coating.

Wash pot *See* Soaking pot.

Waste-heat boiler A steam boiler into which the flue gases of a steel melting or reheating furnace are led before passing to the chimney stack.

Wasters (seconds) Tinplate with defects, but not faulty enough to be rejected completely. Cf. primes, menders.

Waste waste Tinplate with more defects than found on wasters.

Water hammer (obs) *See* Tilt hammer.

Water-seal valve A valve used to control air and gas flows at an open-hearth furnace. It consists of a shallow cylinder fitting into a water-filled annulus when the aperture is to be closed and lifted out of it to open the aperture.

Water tuyere *See* Tuyere.

Wearing plates *See* Wear plates.

Wear plates (wearing plates) Renewable pieces of wear-resisting iron or steel which are fixed at points of heavy wear (eg in sinter or coke conveyors, in hoppers, and in bunkers). They can be replaced easily and prevent wear of the permanent equipment.

Weathering (1) (seasoning) Leaving iron castings out in the open for several months so that internal stresses could settle out. Now virtually obsolete, heat treatment being used instead.

Weathering (2) Leaving iron ore in piles in the open air for several months, for the rain to wash out soluble sulphates.

Weathering steel A low-alloy steel which forms a protective rust on the surface when exposed to the atmosphere and 'weathers' to a dark brown finish. It needs no painting. The American Cor-Ten steel (also made in Britain) is the best-known example.

Wedge bars *See* Screen bars.

Weigh pan (obs) A rectangular box, open at one end, used to contain a weighed charge for a crucible.

Weighting   Moulds can be held together for casting by means of weights placed on the cope. In mechanized foundries the weighting is often done mechanically — a machine places the weights in position and another machine removes them after casting.

Weir   *See* Skim.

Weld   To join two pieces of metal by fusion.

Welding heat   *See* White heat.

Well (1)   *See* Crucible (2).

Well (2)   The area around the nozzle in a bottom-pour ladle.

Wellman gas machine   A proprietary gas producer based on the Siemens producer.

Welshman (obs)   *See* Ring and wedge.

Welsh mill   *See* Jump mill.

Wet-bottom pit   A soaking pit with a silica sand or brick floor. This combines with scale from the ingots and a liquid slag is formed. It is run off at intervals. Cf. dry-bottom pit.

Wet puddling (obs)   *See* Puddling.

Wet regulator   *See* Regulator.

WG   *See* Wire gauge.

Wheelabrator (airless shotblasting)   A proprietary shotblast machine for cleaning castings. The shot is allowed to fall on to one or more rapidly revolving wheels, which throw it off by centrifugal force on to the castings suspended in a sealed chamber.

Wheel rim sections   Steel sections of broadly shallow channel form but specially shaped and rolled for the makers of car and commercial vehicle wheels.

Wheel swarf (obs)   Sand and finely divided metal particles produced during grinding. It was formerly used to form an airtight seal for the lids of cementation boxes. It vitrified on heating and so sealed the joint.

Whelp   A refractory brick of the same thickness and breadth as a standard square (qv) but longer.

Whiteheart iron (European malleable iron)   White cast iron packed with hematite ore in pots and heated to a temperature of about 900°C for several days. Most of the graphite, which causes weakness in ordinary cast iron, diffuses to the surface and is oxidized by the ore, and the iron becomes strong and ductile. It is called whiteheart from the whitish appearance of its fracture. Modern methods use a special gas atmosphere in the furnace instead of an oxidizing ore.

Cf. blackheart iron, pearlitic malleable iron.

White heat (snowball heat) (welding heat) (running heat) (greasy heat) A term used by the smith to describe the heat to which wrought iron or steel must be brought to weld it. It is quite critical but is judged by eye. For wrought iron or mild steel it is about 1300°C.

White iron (1) (obs) Cast iron low in carbon and silicon, especially the latter, made for use in the dry puddling process. It could be produced in the blast furnace but was usually a product of the refinery (qv).

White iron (2) Cast iron in which almost all the carbon is present in chemical combination with the iron and not as free graphite flakes. It is hard, brittle, and difficult to machine, so is generally used only as the raw material for malleable iron castings (qv). Formerly it was used as the raw material for dry puddling (qv).

White iron (3) (obs) A long-obsolete term for tinplate.

White metal A tin alloy used for lining rolling mill bearings.

White pickling Pickling (qv) of a finish-rolled sheet prior to tinning or terne coating.

White pots (obs) Clay crucibles as distinct from plumbago ones.

White rust A form of corrosion found on zinc-coated articles.

White work (obs) *See* Red work.

Whitwell stove (obs) A regenerative blast furnace hot-blast stove, named after the designer.

Wiberg-Soderfors process A direct reduction process, making sponge iron.

Wicket (1) *See* Tuyere cap.

Wicket (2) A small door in a furnace through which the interior can be observed without opening the main door.

Wide-end-up An ingot mould is tapered internally for ease of stripping. It may be cast either wide end up or narrow end up.

Wide strip Flat, thin steel more than about 24 in wide rolled in continuous mills and coiled after rolling into coils which can weigh 10 to 15 tons. The coils can go for further processing in continuous lines, eg galvanizing, tinning; or they can be cut up automatically into sheet lengths. The dividing line is sometimes said to be 18 in, but 24 in is more generally accepted.

Width of corrugated sheets   *See* Pitch.

Wild steel (rising steel)   Steel which has not been deoxidized and gives off gases violently after teeming.

Wilson producer (obs)   A gas producer similar in principle to the Siemens type, but circular and with cleaning holes at the bottom for removal of clinker.

Wind (n) (blast)   The air supplied to a cupola, blast furnace, or Bessemer converter.

Wind box (1)   The space below the tuyeres in a Bessemer converter.

Wind box (2) (blast box)   The pipe round a cupola which conveys the blast or wind from the blast main to the tuyeres.

Wind box (3) (suction box)   The suction chamber beneath the strand of a sinter plant.

Wind furnace (obs)   An old term for any furnace which relied on natural draught instead of a forced air blast.

Wind hop (obs)   An air hoist for blast furnace raw materials, the air being supplied by the blast engine. Cf. tommy hop.

Window   A space in the side of a rolling mill housing, in which the chocks carrying the roll bearings are located.

Wind wall (obs) (spirit plate)   In a square blast furnace the inner wall opposite the tuyere wall.

Winslow's squeezer (obs)   *See* Squeezer.

Wire   *See* List.

Wire drawer's plate   *See* Die (2).

Wire gauge (WG)   Often an abbreviation for British Standard Wire Gauge (qv) but widely used in the nineteenth century to mean all sorts of unspecified wire gauges. In the absence of more detailed information (which is often lacking) it is virtually impossible to tell what WG or wire gauge alone means.

Wire iron (obs)   Very small round iron rod $\frac{1}{4}$ in diameter or even less, rolled in a guide mill for wire drawing.

Wire rod   Small round iron (obs) or steel rods rolled specially

Wire rolling

for drawing into wire.

Wire rolling   Passing drawn wire through flat rolls to produce flattened wire, or through shaped rolls to make shapes of special cross section.

Wobbler (wabbler)   The cruciform end of a roll on to which the wobbler box fits loosely to form a rough universal joint. The wobbler box is also sometimes called a wobbler but is more correctly known as a coupling box.

Wobbler box   *See* Wobbler.

Wolfram   *See* Tungsten.

Wood coal (obs)   A long-obsolete name for charcoal.

Wood hematite   *See* Hematite.

Woodward's cupola (obs)   A cupola in which the air blast was provided by a steam jet in a tube or chimney connected to the cupola stack. The draught was thus induced instead of forced, or blown in, as in the ordinary cupola.

Wool, slag   *See* Slag wool.

Wootz steel (obs)   An old form of steel made in India from wrought iron.

WORCRA continuous steelmaking process   *See* Continuous steelmaking.

Work hardening   Steel, like other metals, hardens and increases in strength, if it is subjected to mechanical work at ambient temperature, eg cold rolling. This can reach a point where it becomes too brittle for use and it must be heat treated. A certain amount of work hardening is deliberately allowed to remain in some cases, as in wire and cold-reduced sheet for applications where stiffness is required.

Working door   The door giving access to a furnace, through which the materials are charged and withdrawn.

Work rolls   The actual working rolls in a four-high or cluster mill.

Cf. back-up rolls.

Worm holes   Cavities rather like worm holes running into a casting usually at right angles to the surface and caused by escaping gases.

Wortle plate (obs)   *See* Die (2).

Wrecking   Removal of the refractory lining of a furnace preparatory to relining. The old lining has no value and is smashed up by various means to get it out quickly.

Wrot iron (obs)   An old spelling of wrought iron (qv) very common in the nineteenth century.

Wrought   A general term for metal which has been shaped by forging, rolling, or drawing.

Wrought iron (obs) (puddled iron)   A commercially pure form of iron with threads of slag or cinder. Can be forged, rolled, drawn, and fire welded but not cast. Was made in the past in various ways and is the oldest form of iron made by man.

Wrought iron grades (obs)   In more recent years British Standard Specifications have been used for grading wrought iron. In falling order of elongation the grades are, Best Yorkshire, Grade A, B, C, D. Prior to the introduction of BSS the grading was by trade terminology; *See* Best iron, Best best iron, etc.

X

X-ray   The familiar X-ray is used in steelworks for non-destructive testing and examination of steel. More powerful examination is possible by gamma rays from radio-active sources. X-rays are also used in special devices for thickness gauging.

Y

Yield   A measure of the efficiency of a furnace, mill, or other process. It equals the actual output divided by the theoretical (ie if there were no waste) and is therefore always less than 100%. In a blast furnace the yield is often called the make.

Yorkshire iron, best   *See* Best Yorkshire iron.

Young iron (obs)   *See* Ready iron.

Z

Zed *See* Z section.

Zinc   A metallic element, symbol Zn, used as an anti-corrosion coating for iron and steel.

Zincote   A proprietary (British Steel Corporation) form of zinc-coated steel supplied in sheet or coil.

Zintec   A proprietary name (British Steel Corporation) for electrolytically zinc-coated steel sheet.

Zircon (or zirconium) paint (or wash)   A refractory coating, painted or sprayed on to cores and mould surfaces as a dressing in steel foundries.

Zirconium   A metallic element, symbol Zr, sometimes used in alloy steels.

Z-mill   A common name for a Sendzimir cluster mill (qv).

Zones   Several types of furnace, especially those for reheating and for heat treatment, are divided for heating purposes into zones, each of which is at a temperature which is carefully controlled, often automatically. So furnaces can be described as triple-zone, five-zone, etc.

Z section (zed) (space bar)   A rolled section shaped like a letter Z, though the angles at the corners are right angles. They are made in steel and were formerly made in wrought iron.

APPENDIX A

Birmingham Gauge
(BG)

Gauge Number	Dimensions		Gauge Number	Dimensions	
	in	mm		in	mm
8/0	0.7083	18.0	23	0.0278	0.707
7/0	0.6666	16.93	24	0.0248	0.629
6/0	0.625	15.875	25	0.0220	0.560
5/0	0.5883	14.94	26	0.0196	0.498
4/0	0.5416	13.76	27	0.0174	0.4432
3/0	0.500	12.7	28	0.0156	0.3969
2/0	0.4452	11.31	29	0.0139	0.3531
1/0	0.3964	10.07	30	0.0123	0.3124
1	0.3532	8.971	31	0.0110	0.2794
2	0.3147	7.993	32	0.0098	0.2489
3	0.2804	7.122	33	0.0087	0.2210
4	0.250	6.35	34	0.0077	0.1956
5	0.2225	5.651	35	0.0069	0.1753
6	0.1981	5.032	36	0.0061	0.1549
7	0.1764	4.48	37	0.0054	0.137
8	0.1570	3.988	38	0.0048	0.122
9	0.1398	3.55	39	0.0043	0.109
10	0.1250	3.175	40	0.00386	0.098
11	0.1113	2.827	41	0.00343	0.087
12	0.0991	2.517	42	0.00306	0.078
13	0.0882	2.24	43	0.00272	0.069
14	0.0785	1.994	44	0.00242	0.0615
15	0.0699	1.775	45	0.00215	0.0546
16	0.0625	1.587	46	0.00192	0.0488
17	0.0556	1.412	47	0.00170	0.0432
18	0.0495	1.257	48	0.00152	0.0386
19	0.0440	1.118	49	0.00135	0.0343
20	0.0392	0.996	50	0.00120	0.0305
21	0.0349	0.886	51	0.00107	0.0272
22	0.0312	0.794	52	0.00095	0.0241

APPENDIX B

British (Imperial) Standard Wire Gauge

SWG No	Diameter in	Diameter mm	SWG No	Diameter in	Diameter mm	SWG No	Diameter in	Diameter mm
7/0	0.500	12.70	13	0.092	2.34	32	0.0108	0.274
6/0	0.464	11.79	14	0.080	2.03	33	0.0100	0.254
5/0	0.432	10.97	15	0.072	1.83	34	0.0092	0.234
4/0	0.400	10.16	16	0.064	1.63	35	0.0084	0.213
3/0	0.372	9.45	17	0.056	1.42	36	0.0076	0.193
2/0	0.348	8.84	18	0.048	1.22	37	0.0068	0.173
1/0	0.324	8.23	19	0.040	1.016	38	0.0060	0.152
1	0.300	7.62	20	0.036	0.914	39	0.0052	0.132
2	0.276	7.01	21	0.032	0.813	40	0.0048	0.122
3	0.252	6.40	22	0.028	0.711	41	0.0044	0.112
4	0.232	5.89	23	0.024	0.610	42	0.0040	0.102
5	0.212	5.38	24	0.022	0.559	43	0.0036	0.0914
6	0.192	4.88	25	0.020	0.508	44	0.0032	0.0813
7	0.176	4.47	26	0.018	0.457	45	0.0028	0.0711
8	0.160	4.06	27	0.0164	0.417	46	0.0024	0.0610
9	0.144	3.66	28	0.0148	0.376	47	0.0020	0.0508
10	0.128	3.25	29	0.0136	0.345	48	0.0016	0.0406
11	0.116	2.95	30	0.0124	0.315	49	0.0012	0.0305
12	0.104	2.64	31	0.0116	0.295	50	0.0010	0.0254

APPENDIX C

Tinplate denominations and sizes on the basis box standard

Mark	Size	Sheets per Box	Weight per Box	Thickness of Sheets	
	in		*lb*	*mm*	*in*
IC	14 × 10	225	108	0.313	0.0123
IX	"	"	136	0.395	0.0155
IXX	"	"	156	0.453	0.0179
IXXX	"	"	176	0.511	0.0201
IC	20 × 14	112	108	0.315	0.0123
ICL	"	"	100	0.292	0.0114
ICL	"	"	95	0.277	0.0109
ICL	"	"	90	0.262	0.0103
ICL	"	"	85	0.248	0.0097
ICL	"	"	80	0.233	0.0091
IX	"	"	136	0.396	0.0155
IXX	"	"	156	0.455	0.0179
IXXX	"	"	176	0.513	0.0201
IXXXX	"	"	196	0.571	0.0223
IC	28 × 20	"	216	0.315	0.0124
IX	"	"	272	0.396	0.0156
IC	"	56	108	0.315	0.0123
IX	"	"	136	0.396	0.0155
IC	20 × 10	225	154	0.313	0.0123
IX	"	"	194	0.394	0.0155
IC	14 × 18¾	124	110	0.309	0.0122
IC	14 × 19¼	120	110	0.311	0.0122
IC	30 × 21	112	243	0.315	0.0124
CL	"	"	224	0.290	0.0114
CLL	"	"	190	0.246	0.0097
CLLL	"	"	176	0.228	0.0090
CLLLL	"	"	160	0.207	0.0081
DC	17 × 12½	100	94	0.404	0.0160
DX	"	"	122	0.525	0.0206
DXX	"	"	143	0.615	0.0242
DXXX	"	"	164	0.706	0.0278
DXXXX	"	"	185	0.796	0.0313